The Manufacture of Iron in Ancient Colchis

David A. Khakhutaishvili

BAR International Series 1905
2009

Published in 2016 by
BAR Publishing, Oxford

BAR International Series 1905

The Manufacture of Iron in Ancient Colchis

ISBN 978 1 4073 0389 5

BAR Publishing is the trading name of British Archaeological Reports (Oxford) Ltd. British Archaeological Reports was first incorporated in 1974 to publish the BAR Series, International and British. In 1992 Hadrian Books Ltd became part of the BAR group. This volume was originally published by Archaeopress in conjunction with British Archaeological Reports (Oxford) Ltd / Hadrian Books Ltd, the Series principal publisher, in 2009. This present volume is published by BAR Publishing, 2016.

Printed in England

BAR
PUBLISHING

BAR titles are available from:

BAR Publishing
122 Banbury Rd, Oxford, OX2 7BP, UK
EMAIL info@barpublishing.com
PHONE +44 (0)1865 310431
FAX +44 (0)1865 316916
www.barpublishing.com

ხახუტაიშვილი დავით ახმედის ძე

რკინის წარმოება ძველ კოლხეთში

CONTENTS

Foreword 1

Researchers have long struggled to make any real sense of how, when and why the manufacture of iron from its ores began and by what stages it developed, and most particularly the way in which steel fits into this problematical picture of early ferrous exploitation. By the mid 20th century a combination a few early written sources, backed up by scanty archaeological evidence - mainly iron artefacts from early contexts - indicated that the original exploitation of iron on any scale most likely began somewhere in the region of eastern Anatolia and the Caucasus, sometime during the first half of the second millennium BC. Even since the mid 20th century the overall lack of evidence has inevitably meant that any sort of cohesive picture has been very difficult to achieve, although various ideas have been put forward, the best known of these by a series of Western scholars.

The best overall survey of the evidence to date is given in Radomir Pleiner's *Iron in Archaeology* (Pleiner, 2000), but even in this the Georgian evidence is barely mentioned. Since World War II the situation has been further hampered by the comparative isolation of the countries which were governed as part of the Soviet Union, and the difficulty (in the West) in being able to read the languages in which relevant research work was published. Thus it is the aim of the present publication of the English translation of David Khakhutaishvili's book *The Manufacture of iron in Ancient Colchis*, originally published in Russian in 1987, to make the results of archaeological research on early iron exploitation carried out in Georgia over a 25 year period (1960-84) - almost unknown in the West - much more widely available.

Early iron working sites in this area south-east of the Black Sea, bordering the Caucasus mountains to the north and eastern Anatolia to the south, also only really began to be identified and investigated, and their significance recognised, after the end of World War II. The results reported here cover the long campaign of excavations, to explore the early iron smelting industry in this region, that began in 1960. By the end of this campaign approximately 400 prehistoric iron smelting sites - mostly clustering in four principal areas, centred on a series of river systems - had been identified. The results of the excavation of 26 of these sites, some from each of these four main areas, form the subject of this book.

Each of the principal areas is identified by the river system in which it was found, and the sites from these are discussed sequentially working from south to north, starting with the Chorokhi [1] river valley (3 sites/6 furnaces), and followed in turn by those of the Cholokhi-Ochkhamuri [2] (8 sites/11 furnaces), Supsa-Gubazeuli [3] (13 sites/15 furnaces) and [4] Khobi-Ochkhomuri (3 sites/3 furnaces) river valleys. The iron ore exploited in the first three of these areas consists of magnetite rich sands, whereas haematite ores were being used in the fourth area, which was also the furthest inland, and nearest the Caucasus mountains to the north.

As the author makes clear the present volume leaves various problems to do with this industry still to be resolved and it is clear that it is not intended to be a definitive publication on the early iron exploitation of this region, which remains to be done and published as a more complete volume *Iron Manufacture in Ancient Colchis*, and it is hoped that this might be possible at some stage. In particular further work is needed on the dating of the various sites but the work so far gives a broadly consistent picture of the early exploitation of iron in the Georgian region. This indicates that three of these principal areas ([1], [2] and [4]) were operating between about the early 1st millennium and about 600 BC, give or take about a hundred years or so. The remaining principal region [3] - centred on the Supsa-Gubazeuli river system in central western Georgia - which is reported as having the richest ores, appears to have been operating for much longer, with a series of archaeomagnetic and radiocarbon dates indicating that iron smelting in this region is active from roughly 1800BC to about 600 BC. One tentative suggestion that can be made is that the Supsa-Gubazeuli iron smelting industry is a relatively localised, very early iron production centre which became much more widespread in this region by the end of the first millennium BC, but much more research would be needed to explore this possibility.

It is only possible to get a very rough idea of the eventual iron output of the 400 or so prehistoric iron smelting sites so far known for Georgia based on the 35 furnaces described here, although a total volume of 146 cubic metres of slag are reported for 22 of the furnaces, giving a mean slag output of 6.6 cubic metres, with an approximate weight of 10 tonnes (assuming slag of varying density etc). The results from the sites investigated so far would suggest that about 700 furnaces can be expected for the 400 iron smelting sites identified up to 1987, indicating a total slag output of approximately 7000 tonnes. Experimental work (by Peter Crew, 1997) would suggest that a 2:1 ratio by weight of slag to (consolidated bloomery) iron might be expected, which would suggest that a very rough estimated output of 3,500 tonnes of iron could be made for the prehistoric Georgian iron smelting sites known at this time.

However the true figure may be much higher, given that there may well be many more than 400 sites/700 furnaces in total. David Khakhutaishvili has himself said that there is much more work to be done on this industry, and the more recent finding of another principal iron smelting area - this time centred on the Chakvistskali river valley, north of Batumi (reported by Nana Khakhutaishvili in 2005) supports this view. The results reported in this (the 1987) volume indicated that the peak period when most iron was produced in prehistoric Georgia was during a roughly 300 year period from the 10th to the 7th century BC. Using the estimates given above this could represent a very approximate annual output of 10 (+?) tonnes during this period.

The rough estimates of iron output are only suggested here to give a rough idea of possible yields at this time although the (known) survival of actual ironwork of this period (as at other times) is extremely uneven, and dependent on how and why this material got into the ground just as much as the (often haphazard) geographical evenness of archaeological investigation. However the scale of iron production at this period can be gauged by the enormous hoard of ironwork (?tribute), amounting to about 163 tonnes, found in the palace complex at Khorsabad in the upper reaches of the River Tigris, the neo Assyrian city built for Sargon II (d. 705 BC) and subsequently abandoned. Much of the ironwork in this hoard was in the form of fish-shaped biconical bars of 'trade' iron, which would appear to be part way between a consolidated bloom, and a finished artefact. This may well have been the form in which the iron left the contemporary smelting sites such as those in Georgia even if none have yet been noted. However this form of 'trade' iron has been noted from the Levant in contexts as early as the 9th century BC and must have been a widely used and very long lived form of 'trade' iron as it was still noted in this region - in much the same shape and size - as being found in south-western Iran, in the early 9th century AD, some 1,800 years later.

Most of the early Georgian smelting sites described here are remarkably consistent in their layout with either one or two furnaces (possibly operated as a pair despite the apparent dating contradictions), a nearby low clay and stone built working platform with a hearth for reheating the blooms before these were consolidated on a stone anvil. There was also a nearby charcoal supply (that could have served both the re-heating hearth - the presence of which is indicated by the degree of burning found on this working platform - and the inevitable waste heap/slag dump making use of any nearby hill slope. The remains of hammer stones, mortars and the like for ore crushing were also noted.

The results of the present publication indicate a large, well developed industry, although it is likely to be based on one or other variant of a shaft furnace, the upside down 'bowl' type of furnace superstructure shown here in the reconstruction, having come from an erroneous earlier suggested reconstruction and not based on evidence. Further survey work should give us a better idea of just how large this prehistoric iron industry was, but it is already clear from the results reported in this book that the furnaces varied in size, with some being very large, with an internal diameter (at ground level) of at least a metre across. Also the very consistent way in which the furnace sites appear to have been laid out and used would suggest an accepted or 'standardised' way of operating. The furnaces are of a well built, slag pit type, with the pits being either all or part stone lined, the lining being coated with refractory clay.

It is also concluded that some of the furnaces are likely to have been fired (and perhaps designed too?) with the specific intention of producing a steely bloom. This is an obvious way in which bloomery smelting may have developed from quite early on (during the 2nd millennium BC) although we cannot (so far) see this directly from the furnace remains, and associated slag waste material reported here. This possibility is further developed here by looking at the possibility that the name 'Chalybes' (makers of iron) might actually have derived - of have come in the first place - from a more specific early meaning (makers of good iron), possibly referring to them being the makers of steel as a specialised variant of bloomery iron.

Although this relatively short book can be seen as more of an interim description if the findings from a 25 year campaign of archaeological investigation into the nature of the early iron manufacturing industry in Georgia - as is clear from the unresolved issues which are mentioned in places - its appearance, both in the original Russian language form, and most particularly in this English translation, means that a great deal of otherwise virtually unknown evidence on the existence, form, date and scale of this early iron working industry is now accessible, and allow further research to be planned and undertaken. In fact the recent appearance (in 2007) of a short monograph *About the History of Iron Production in Georgia* (by Givi Inanishvili, one of David Khakhutaishvili's co-researchers) - looking partly at the iron production technology, but mainly at the production and metallography of prehistoric (mainly 1st millennium BC) iron artefacts - shows that this is has already begun to happen. It is greatly to be hoped that this will continue to happen, but further research would not have been possible without the results of all the groundwork and post-excavation processing that are presented in David Khakhutaishvili's book.

Brian Gilmour,
Research Laboratory for Archaeology,
University of Oxford

Foreword 2

To many English-speaking readers the archaeology of the Caucasus has until quite recently been a largely closed book, often concealed behind publications which were for the most part difficult to get hold of or languages which few were able to read. Many prehistorians, however, have harboured a conviction that this is an area which, if its archaeology were only more accessible, has the potential in one way or another to shed light on a number of major questions, particularly those concerning various aspects of early metallurgy in western Asia. This important and interesting volume does just that. By systematically presenting in English the results of excavations carried out between 1960 and 1984 on 9 clusters of iron-working sites in western Georgia, it goes a long way towards illuminating the production of iron in this area in the second and early first millennia BC.

Some of the earliest iron objects known archaeologically have been found in Anatolia, at third millennium sites such as Tilmen Höyük, Troy or Alaça Höyük. In the early second millennium iron is mentioned in the Old Assyrian texts from the *karum* at Kultepe-Kanesh, in which it is valued many times more highly than silver; and by the later second millennium the Hittite documents from Bo□az Köy list an impressive variety of iron objects and implements stored in the royal treasuries. Where the Hittite kings - and other contemporary rulers, who delighted in giving gifts of iron knives, daggers *etc.* to one another - got this iron has hitherto never been at all clear, but this book would seem to suggest that there is now at least a strong possibility that some of it was produced by small-scale iron producers in the south-eastern Black Sea region. The dates obtained from over 30 iron-smelting installations, by means of calibrated C^{14} determinations from charcoal found in the furnaces, archaeomagnetic dating (normally reckoned to date the last firing of the furnace) and dating of associated ceramics (or a mixture of these) range between the 19th and 6th centuries BC, with the majority (as perhaps might be expected) clustering in the 9th-7th centuries, which coincides with the height of the nearby Iron Age Kingdom of Urartu and its plethora of iron finds.

The workshops themselves have been identified by the presence of furnaces, iron slag and other waste, bellows nozzles and stone tables suitable for use as anvils. The dates, while not perhaps individually always to be taken too literally (the problems of 'inbuilt age' in C^{14} determinations derived from charcoal are notorious; in at least two instances the archaeomagnetic date [supposedly representing the latest use of the furnace] appears to be earlier than the calibrated C^{14} date; and, in the case of those furnaces which produced the earliest C^{14} dates [Mshvidobauri II, Nagomari I, 2 and Mziani II and IV], the pottery is either undiagnostic or non-existent), are on the whole consistent, and appear to indicate that Colchis was one of the more precocious iron-producing regions in the later second and early first millennia, and may well have been so at an even earlier date. This is perhaps not surprising, since Transcaucasia and the south-eastern Black Sea can also lay convincing claim to have been one of the most productive and innovative bronze-working areas from at least the Early Bronze Age (when its products seem likely to have contributed to the distinctive character of much central and northern Anatolian bronzework), and there can be little doubt that the early production of smelted iron initially grew up hand-in-hand with that of bronze, whether adventitiously or as the result of experimenting with different ores and fluxes. It is surely no coincidence that those areas apparently most forward in the production of iron in the late second and early first millennia - among them western Iran, Cyprus and the south-west Caucasus - were also areas well known for their skills in bronze and other metallurgy. The incentive, given the economic and cultural esteem accorded to iron in the Bronze Age of western Asia and south-east Europe, is not hard to understand.

It is clear from Professor Khakhutaishvili's account of the study of the history of ancient iron manufacture that much effort in the past has been devoted to identifying one area or another or one "people" or another of the ancient world as the primary centre of origin or inventors of iron-working, sometimes from barely concealed political motives. This seems to me about as rational as measuring national standing by Olympic medals or competition in the space race, and it appears to stem from a belief that technological advances in themselves form the most significant milestones of human history. This book, by presenting the results of over twenty years of excavation of metallurgical sites in full archaeological and scientific detail and by discussing them in the light of varied technological and ethnographic data, allows us to move beyond such irrelevancies and to begin to put the study of ancient iron production on a sounder footing within a wider and more balanced perspective.

Susan Sherratt
Department of Archaeology
University of Sheffield

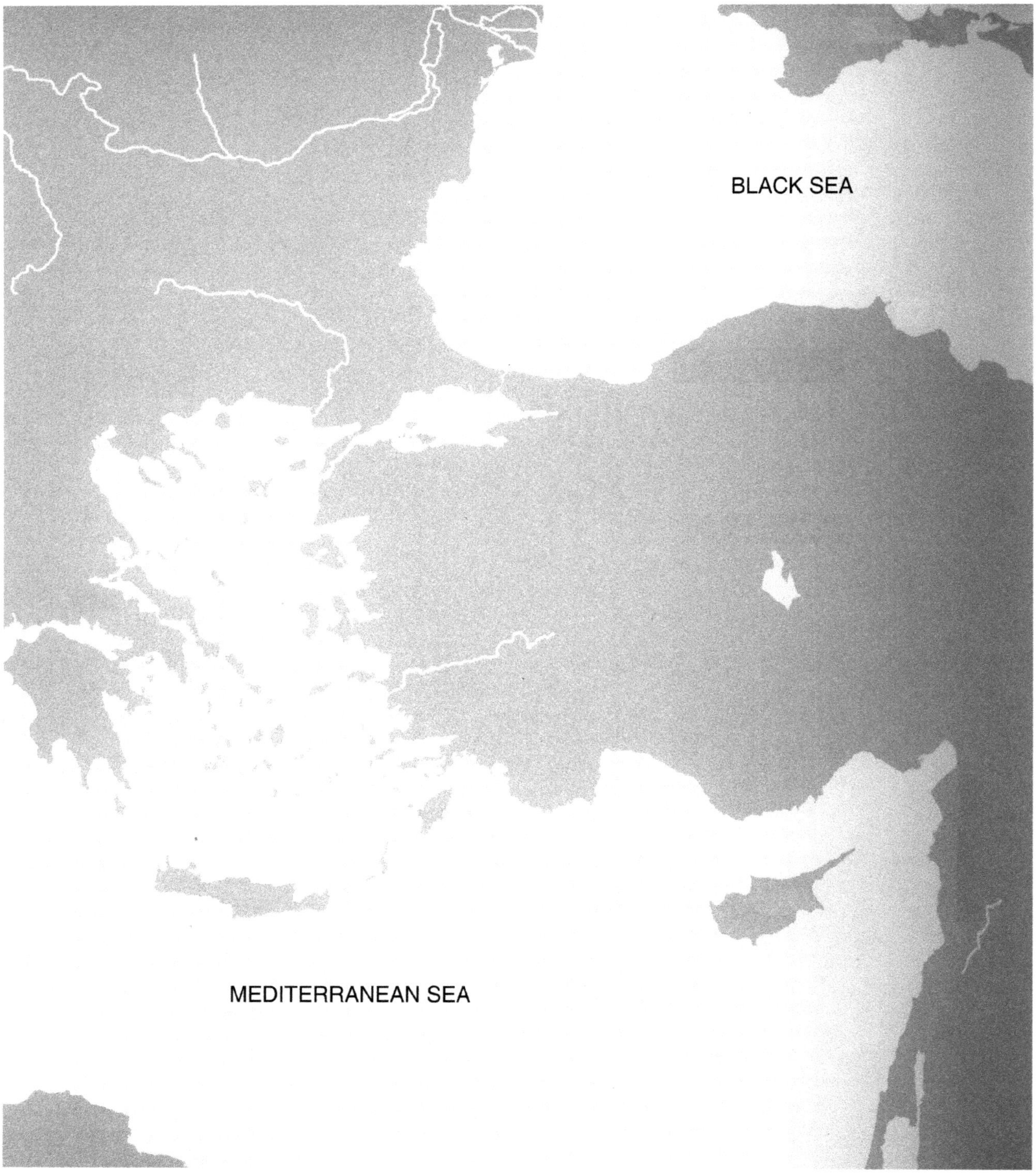
BLACK SEA
MEDITERRANEAN SEA

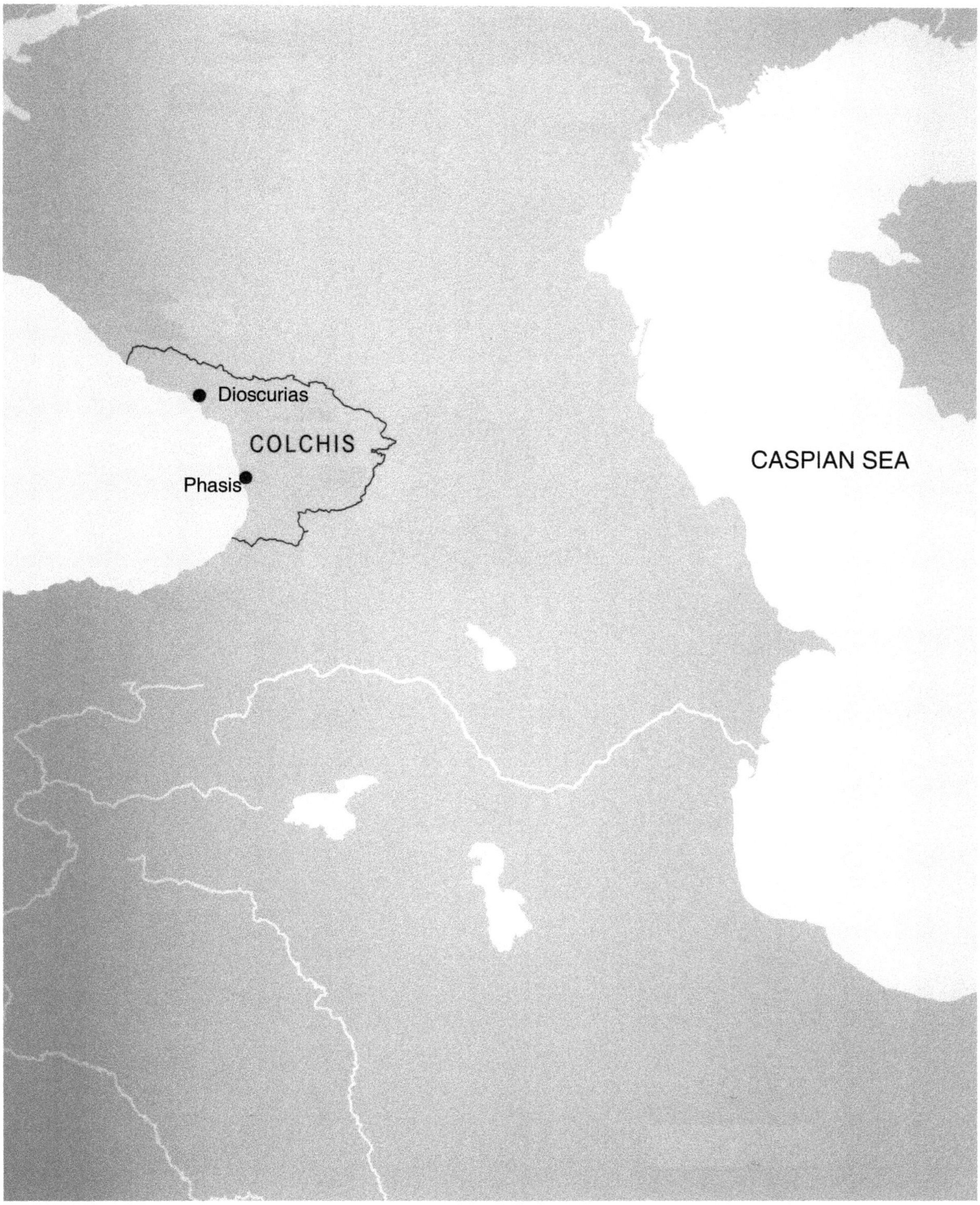

Dioscurias
COLCHIS
Phasis
CASPIAN SEA

PREFACE

During the last few decades archaeological research in Colchis has achieved signal successes: materials of first-rate scientific value were obtained through the efforts of a number of researchers[1] on the economic, social, political, ethnic and cultural history of the tribes that inhabited the south-eastern Black Sea littoral in ancient times.

Some of the data collected in the process of investigating the history of ancient Colchis are of great importance for research into a number of key problems concerning the past of not only the region just mentioned, but also concerning those of the entire Transcaucasus, the Near East and the Aegean world. One of the monuments that yielded materials of interest to us is the Colchin metal extracting and iron producing area[2]. This area is at present regarded as the most ancient and most productive mining-metallurgical centre of its kind.

Regular excavations of the above group of monuments, a few of which have been mentioned before[3], were launched in 1960 by I.A. Gzelishvili[4].

In 1960-1961, I.A. Gzelishvili excavated six sites of ancient Colchian metallurgy. Ten smelting furnaces were unearthed together with ten dumps of waste and other attributes of iron-smelting workshops[5].

The publication of the preliminary results of the first excavations that had collected abundant factual material radically changed earlier conceptions of the chronology, composition and scientific value of the centre of iron manufacture that interests us. It was found that there existed a single centre of iron production, small sites covering the entire foothill area of the Eastern Black Sea littoral, as well as, most probably, the south eastern part of it too[6]. This necessitated a new complete publication of both the old materials published by I.A. Gzelishvili and the newly obtained ones.

The present book contains data on 26 sites of ancient iron manufacture where excavations brought to light 35 iron-smelting furnaces and 39 dumps of waste, as well as other component parts of iron-smelting workshops.

It should be stressed that excavations of iron-smelting workshops do not always yield material that can be dated by archaeological methods; in those days settlements were, as a rule, situated at a considerable distance from manufacturing sites, while the waste dumps contained only sherds of ceramic vessels which the craftsmen had broken by chance and which were not fit to be restored. Therefore, when determining the date of a workshop, we resorted, whenever possible to radiocarbon and archaeomagnetic data: the thing is that remnants of iron-smelting furnaces are fine samples for determining changes in the parameters of geomagnetic field elements. Besides, the interiors of iron-smelting furnaces were often coated with layers of particles of charcoal which facilitated dating the monument by the radiocarbon method.

All this taken together made it possible to reconstruct the chronology of the excavated objects of ancient iron manufacture, to plan further investigations and to determine the approximate time of their completion.

Concrete problems of the technology of ancient iron manufacture are not considered in the present book: these problems are being studied and will be published by specialists in this branch of science[7].

The work of archaeological expeditions excavating objects of ancient Colchian metallurgy was carried out at various times by I.A. Gzelishvili (1960-1961), A.K. Inaishvili (1960-1961, 1970-1971), D.A. Khakhutaishvili (1960-1984), A.T. Ramishvili (1960-1961), A.Y. Kakhidze (1961), L.B. Chkhaidze (1960-1961), S.I. Gogitidze (1979-1984), A.M. Djavelidze (1970-1984), N.D. Khakhutaishvili (1972, 1974), R.A. Atoyan (1960-1961), V.M. Poplavskaya (1960-1961), V.N. Mitaishvili (1960-1961), and others.

Samples of organic origin were dated by the radiocarbon method in the dating laboratory of Tbilisi State University under the supervision of A.A. Burchuladze and C.I. Togonidze, while the archaeomagnetic method was used in the Centre for Archaeological Research under the auspices of the I.A. Djanasha Institute of History, Archaeology and Ethnology of the Georgian Academy of Sciences, under the supervision of Z.A. Chelidze.

---

[1] A.A. Melikishvili, *On the History of Ancient Georgia*, Tbilisi, 1959. T.K. Mikeladze, *Research into the History of the Most Ancient Population of Colchis and of the South-Eastern Black Sea Littoral*, Tbilisi, 1947 (Text in Georgian, summaries in Russian and English); O.D. Lordkipanidze, *Ancient Colchis, Myth and Archaeology*, Tbilisi, 1979.

[2] This term was suggested by us after excavations carried out in the 1970s on the territory belonging to the following villages: Tsetskhlauri, Askana and Pirveli Choga (Kobuleti, Makharadze and Chkhorotskhu districts, Georgian SSR).

[3] O.M. Djaperidze, "An Exploratory Expedition in Guria," *Annals of the Acad. S.N. Djanashia State Museum of Georgia*, XVI-B, 1950, p. 29. N.Y. Khosharia, "Archaeological Research at Ureki," *Materials and Research into the Archaeology of Georgia and the Caucasus*, I, 1955. p. 29.

[4] I.A. Gzelishvili, "Basic Results of Excavations of Iron-Smelting Workshops in Adjaria," *Monuments of South-Western Georgia*, I, Tbilisi, 1964, pp. 29-44 (text in Georgian); *Iron Smelting in Ancient Georgia*, Tbilisi, 1964, pp. 5, 38-52; D.A. Khakhutaishvili, "The Beginnings of Colchian Iron Metallurgy," *Monuments of South-Western Georgia*, I, pp. 45-58.

[5] Gzelishvili, *Iron Smelting*, pp. 38-52 (text in Russian).

[6] D.A. Khakhutaishvili, "Contribution of the Kartvelian Tribes to the Mastery of Iron Metallurgy in the Near East," *Acta Antiqua*, 22, Fasc.1-4, Budapest, 1974, pp. 339-348.

[7] P.E. Tavedze, G.V. Inanishvili, T.N. Sakvarelidze, T.N. Zague, "Investigation of Ancient Slags of Iron Manufacture of the Territory of Georgia," *History of Science*, Tbilisi, 1984, pp. 20-25.

I consider it my duty to express my gratitude for the assistance accorded to us throughout the years during which we studied the ancient Colchian centre of iron metallurgy; this assistance greatly facilitated both field work and laboratory research.

We must alas admit that we do not yet posses the data required for a statistical analysis of the development of ancient iron manufacture. Such data will become available to specialists only when more objects of ancient Colchian siderurgy are excavated and classified. I consider it expedient to wait for that time; I therefore, offer the present book to the attention of specialists.

This translation has come about through the devotion to her late father's memory of Nana Khakhutaishvili. It was seen through the press by Michael Vickers, who wishes to thank Brian Gilnour and Susan Sherratt for their contributions, and David Davison for having included the title in the BAR International Series.

INTRODUCTION TO THE STUDY OF THE HISTORY OF ANCIENT IRON MANUFACTURE

A large number of works have been published on the history of early iron metallurgy and on the primary centres of origin of siderurgy; a great many hypotheses and suggestions have been put forward; the problem has been considered specially and in passing, individually and collectively. Despite all this, the problem of the primary centre where iron manufacture originated is still far from being solved. Numerous logical argumentats have been proposed at different times and by different outstanding scientists; but this fact merely stressed the urgency and the scientific interest of the problem, adding very little to its solution. One will be convinced of this by merely running through part of the special literature and comparing the propositions suggested in it with the materials presented in this book.

Before we start reviewing relevant publications[1] we should note that, as far as we know, material evidence suitable for the delimitation of concrete regions where iron was manufactured in the second millennium B.C. (iron-smelting furnaces, iron-smelting workshops, well dated heaps of slag, etc.) have been found exclusively in the foothill areas of the eastern Black Sea littoral (Ancient Colchis). Pits from Gerar dated by W.F. Petrie to the 12th to 10th centuries B.C., and re-dated by G.E. Wright to the 10th to 8th centuries B.C., were erroneously taken for remnants of iron-smelting furnaces;[2] in reality, they were pits for household needs that had nothing to do with the manufacture of iron.

Before Petrie's work came out, Blinkenberg stated that iron metallurgy originated in the southern Black Sea littoral where the Hittites knew how to smelt iron and temper it.[3]

A few years later, after the publication of Petrie's work, T.A. Rickard reiterated the opinion that the bloomery hearth method of manufacturing iron was widely used by them as early as the 15th century B.C.[4] Thus, Blinkenberg and Rickard, proceeding from the scanty data concerning iron gleaned by them from Hittite cuneiform texts, considered that Hittite territory was the primary centre where the technology of manufacturing iron in bloomery hearths originated.

[1] The present review does not claim to dwell on all the publications actually existing, seeing that the materials available to the author present a complete picture of the state of things in the study of the problem in question during the last few decades.

[2] W.F. Petrie. *Gerar*. London, 1928, pp,14-16, Later Wright re-dated these pits to the 10th to 8th c. B.C. See G.E. Wright. "Iron: the Date of its Introduction into Common Use in Palestine," *AJA*, 43, (1939) pp. 450-462.

[3] Blinkenberg. "De pays natal du fer". *Mém.Soc.Roy.Antique du Nord*, 1925, pp. 204-2-5.

[4] T.A. Rickard. *Man and Metals*, II, New York, 1932, p. 870.

The above hypothesis, as we shall see, became one of the most popular and widespread, while old ones continued to exist and new ones appeared.

Inasmuch as antique sources regard the Chalybes as inventors of iron, B.E. Deguin-Kovalevsky opines that the possibility of Caucasian iron being exported to the countries of the Mediterranean littoral cannot be excluded.[5] Having considered sources known to science in the 1930's, he came to the conclusion that there existed two areas of iron manufacture in the Transcaucasus: the eastern area - Kakhetian-Armenian, and the western - Imeretian-Megrelian-Abkhazian. His deductions are based upon geological data (presence of raw material), ethnology and folklore (cult of iron), iron manufacture by peasants which continued almost to our day and upon certain indirect archaeological data.[6] Careful consideration of Deguin-Kovalevsky's argument reveals it to be unconvincing and somewhat naive; but, taking into consideration the fact that the Transcaucasus was insufficiently studied from the archaeological point of view, we have to give him his due for his perspicacity and breadth of vision.

A few years after the publication of B.E. Deguin-Kovalevsky's work, S. Przeworski voiced the opinion, proceeding basically from archaeological data, that in the 13th and 12th centuries B.C. iron was a rare metal in Asia Minor, while in the Transcaucasus it appeared still later. According to him, Asia Minor could not be regarded as an early centre of iron manufacture for the Near East, as it was, rather, closely connected with other neighbouring iron-manufacturing regions and developed parallel to them.[7]

B.A. Kuftin stated, in a monograph that came out in 1941, that iron metallurgy was a secondary phenomenon in the Transcaucasus (in the Caucasus) and the place of its origin was Nearer Asia.[8] Subsequently his view was supported by B.B. Piotrovsky,[9] E.I. Krupnov,[10] A.A. Martirosyan,[11] O.A. Danelyan[12] and several other researchers.

[5] B.E. Deguin-Kovalevsky. *On the History of Iron Manufacture in the Transcaucasus* 120, Moscow Leningrad, 1935 p. 242.

[6] Ibid., p. 242 ff.

[7] S. Przeworski. "Die Metallindustrie Anatoliens in der Zeit von 1500 bis 700 v.Chr". *Internationales Archiv für Ethnologie*. Vol. 36, Leiden, 1930, p. 428.

[8] B.A. Kuftin. *Archaeological Excavations in Trialeti*. Tbilisi 1941, pp. 70-72.

[9] B.B. Piotrovsky. *Archaeology in the Transcaucasus*. Leningrad. 1949, pp. 115-116.

[10] E.I. Krupnov. *Ancient History of the Northern* Caucasus, Moscow, 1960, pp. 8-9.

[11] A.A. Martirosyan. *Armenia in the Epoch of Bronze and Early Iron*. Yerevan, 1964, pp. 131-132.

[12] A.A. Danielyan. The Culture of Early Iron on the Territory of Azerbaijan. Summary of master's thesis. Baku, 1969, pp. 13-15.

As G.A. Melikishvili put it, "quite possibly, the highly developed iron metallurgy of Asia Minor and of the south-eastern Black Sea littoral influenced the progress of iron manufacture of both the Transcaucasus (Western Transcaucasus in particular, whose culture was since antiquity in close contact with both the south-eastern Black Sea littoral) and with Urartu itself."[13]

In another publication Melikishvili remarked that the Hittite term "kunna" meaning "copper" as well as the ancient Greek κύανος, which is considered to be a loan word, may be likened to the Georgian term "rqina." In the Zan language this term occurs in the form "rqina-qina" (Megrelian) and "erqina" (Chanian). This is comparable with the Armenian "erqat" ("iron") which is thought to have been borrowed from the Kartvelian languages. According to Melikishvili, "If we accept the interpretation of the relevant Hittite term suggested by A. Götze, the term "rqina" used in the Georgian (i.e., Georgian-Megrelo-Chanian) milieu should be rather regarded as a later borrowing from the Hittite-Asia Minor world... This may be testified to by the fact that this term is known to only two or three Kartvelian languages - to Georgian and Megrelo-Chanian, while it does not exist in Svanian, a language that branched off from the parent common Kartvelian at a much earlier date. Consequently, this term did not exist in the Kartvelian parent language, i.e. at the turn of the third and second millennia B.C."[14]

Thus, according to Melikishvili Asia Minor may be regarded as the primary centre of origin of iron metallurgy in the Hittite epoch.

A. Lukas considered (1948) that western Asia was the original centre of mining and manufacturing iron.[15]

According to R.J. Forbes who could have had at his disposal only a few data concerning iron provided by Hittite cuneiform texts, i.e. data on smelting spongy iron, it must have originated somewhere in the mountains of Armenia between the Taurus and the Caucasus. From the 19th to the 14th centuries B.C., iron was used for inlay work and for making expensive ceremonial weapons. He admits that certain knowledge of iron smelting may have spread from this region to others. Earlier attempts to obtain iron preceding the 19th century B.C. were most probably too irregular; Forbes supposes that about 1400 B.C. the Chalybes, subjugated by the Hittites,[16] discovered the technique of "steeling", which gave the Hittites a lease of two hundred years of monopoly for the production of "good iron" or steel. Since then iron artifacts appear more and more often as objects of trade. In the 13th century B.C. they occur even as far as in northern Europe; still, iron remained for a long time a secondary metal as compared to copper and bronze.[17]

Forbes asserts that the invasion of Asia Minor by "maritime peoples" about the 12th century B.C. and the subsequent fall of the Hittite state furthered the spread of knowledge of the technology of smelting and cementing iron. Between 1200 and 1000 B.C. there is a boom of iron manufacture in Iran, the Transcaucasus, Syria, Palestine, in Cyprus, Crete, in Mesopotamia, and in the Caucasus.[18]

As Forbes put it, "the history of iron manufacture in Armenia and in the Caucasus is insufficiently studied for the lack of reliable information... It is unknown whether the aborigines of Urartu, Herodotus's Alarodians had any contacts with the iron mining or "metal culture" of Pontus... The arrival of the Age of Iron in Mesopotamia is certainly due to the intense extraction of iron in the mountains... Several ornaments of iron dating from the 13th century B.C. were found in the Gandja-Karabakh region, while during the subsequent two centuries there were iron weapons repeating the shape of weapons in bronze...; at the same time, iron appeared in Georgia and Armenia[19] ..., but the Age of Iron had finally won in the Scythian period.[20]

Touching upon the role of the Chalybes in the formation and development of iron metallurgy and mentioning the opinion of Geihelheim that the cementation of iron and bringing it to a quality surpassing bronze was discovered by the Chalybes-Chaldes on the Pontic littoral under the influence of Hittite masters, Forbes adds his own part, that the cementation of iron smelted by man originated in the Hittite region, probably in the mountains of Armenia but not in Europe. However, he remarks that the problem of the exact location of the area where the earliest "steel manufacture" originated can be solved only on the basis of further research. He does not exclude the possibility that the region where "iron is born" may turn out to be Dolikhé, or it may be corroborated that the Chalybes were really aborigines and handled iron as masterfully as they did other metals.[21]

Considering the Accadised Hittite word "hapalkinnu" – "iron", Forbes remarks that this word brings us to the problem of the Chalybes, the "tribe of blacksmiths", "siderurgists", who inhabited the Black Sea coast between Samsun and Trebizond (Trabzon). According to him, no matter of what ethnic origin the Chalybes were, the Pontic region played an important part in the history of iron metallurgy long before the 12th century B.C. when the Chalybes fell under the domination of the

[13] Melikishvili. *On the History of Ancient Georgia*. Tbilisi, 1950, pp. 200.

[14] Melikishvili. "On Some Names of Metals in Ancient Oriental and Caucasian Languages". *Bulletin of Ancient History*, 1968, 4, pp. 124-125; ibid. references.

[15] A. Lukas. *Materials and Handicrafts in Ancient Egypt*. Moscow, 1958, p. 371.

[16] It should be noted that Forbes' surmise is unconvincing since the Chalybes who dwelt in Asia Minor in the first millennium B.C. were not "subjugated by the Hittites" during the existence of the Hittite state.

[17] R.J. Forbes. *Metallurgy in Antiquity*. Leiden, 1950, p. 418; *A History of Technology*. Edited by Ch. Singer, E.J. Holmayer and R. Holl, vol. 1. 1955, pp. 592-595.

[18] Ibid., p. 419.

[19] Ibid., p. 449.

[20] Ibid., p. 450.

[21] Ibid., pp. 420-421.

Musei and Tibarenes. He holds that the Chalybes had come to the Pontic coast from another area, but after 1200 B.C. some of them moved to various regions, thus furthering the propagation of knowledge of iron manufacturing techniques, the monopoly of which was lost after the fall of the Hittite state.[22]

Summing up his investigation of the gradual development of iron manufacture, Forbes suggests singling out three stages:

I. The use of meteorite iron;
II. Obtaining iron as a secondary product from purifying gold;
III. Reduction of iron ores (after long experimentation) in bloomery hearths.

This latter achievement came to play a leading part particularly after the discovery of cementation methods and solved the problem of producing steel, thus ushering in the Iron Age;[23] but difficulties of slagging smelted ore, smithing the bloom, heating it white-hot, as well as the complex processes of tempering postponed the actual inception of the Iron Age[24] which set in only after the processes of carbonization (cementation), cooling, and tempering had been mastered one after another. As Forbes states, analyses of iron objects from Egypt corroborated that from 1200 B.C. to 900 B.C. iron objects were only carbonized between 900 and 700 B.C.; the technique of annealing and tempering was introduced only in Roman times.[25]

Forbes lacked a very clear idea of the ethnic geography of the Caucasus and the south-eastern Black Sea littoral, although, for his time, he had a wealth of information which, unfortunately, contained no description of at least one Near-Eastern pre-antique iron-smelting furnace: in those days science did not have at its disposal any reliable data on early iron-manufacturing centres, nor on the technical means used by the first master-smelters of iron.[26]

In 1956, having analyzed the materials that were available to Forbes, H. Coghlan came to the conclusion that the history of iron manufacture cannot be made sufficiently clear until new data are provided.[27]

A few years later, L. Aitchison attempted to substantiate the theses formulated by other researchers that the primary centre of iron manufacture was Chalybia, one of the provinces of the Hittite empire, located on the Black Sea littoral, in the basin of the Halys River. Aitchison opines that the Hittites held a monopoly of the secrets of iron manufacture and it was only from the 15th century B.C. that the know-how of smelting iron ore began to spread.[28]

According to S. Foltiny, Asia Minor was the primary centre of iron manufacture and the place of its origin. Since antiquity, Anatolia had a powerful influence on the Aegean world and farther westward, including Italy where bronze articles from Asia Minor were widespread owing to trade. Technological knowledge of iron smelting followed the same path, from East to West.[29]

In 1961, R.M. Abramishvili summed up the work he had done on the history of mastering iron manufacture in Georgia. He surmised that iron came to be used on the territory of Eastern Georgia, especially in the region called Kvemo Kartli (Lower Kartli). The material on which this deduction was based was iron objects found mainly in burial complexes In Kvemo Kartli and in Mtskheta.[30] The earliest date, namely the 14th century B.C., was assigned by him to a burial complex consisting of a cist grave unearthed near the village of Beshtasheni, on the right bank of the Baiburt river, and containing, among other things, an iron dagger with a bronze hilt.[31] Abramishvili assigned almost the same date, the end of the 14th or the 13th century B.C., to bronze buckles inlaid with iron from Kobani and Abkhazia.[32]

Pit burial No. 13, from Beshtasheni (where among the grave goods an iron knife blade[33] was found) was assigned by Abramishvili to the turn of the 14th and 13th centuries B.C. The 13th century B.C. or the turn of the 13th and 12th centuries B.C. was the date assigned to the grave goods from pit burial No. 18 from Beshtasheni containing an iron dagger blade.[34] A hoard from the village of Ude (Aspindza district, Georgian SSR) containing, besides numerous objects of bronze, two iron spearheads, a dagger blade, a mace and a pin were dated to the turn of the 13th and 12th centuries B.C.[35] R.M. Abramishvili assigned to the epoch of Early Iron, specifically to the turn of the 12th and the 11th centuries B.C., a long iron knife blade from burial No. 6 and an iron dagger from burial No. 60 near the village of Tsinskaro; also iron objects including an iron knife from

[22] Ibid., pp. 453-455, cf.p. 280. *A History of Technology*, p. 595
[23] Ibid., p. 414.
[24] Ibid., p. 417.
[25] *A History of Technology* I, 1955, p. 596.
[26] R.Forbes is right in recalling that "little can be said about the evolution of smelting furnaces and hearths. Documentary evidence is scarce and is not in agreement, and further detailed investigation is imperative." *A History of Technology*, I, p. 598.
[27] H.Coghlan. *Notes of Pre-Historic and Early Iron in the Old World.* Oxford, 1956, p. 70.

[28] L. Aitchison. *A History of Metals*. I, London, 1960, p. 101. This was, for its time, the first two-volume monograph on the history of metals, from antiquity to our day; a critical review of it was published by Y.A. Pozukhin *Sov. Arch.*, 1963 No. 3, pp. 265-271.
[29] S. Foltiny. "Athens and the East Hallstatt Region". *ASA*, 65, 1961, pp. 283-297.
[30] R.M. Abramishvili. "On the Mastering of Iron Manufacture on the Territory of Eastern Georgia," *Annals of the Acad. S.N. Djanashia State Museum of Georgia*, 22B, Tbilisi, 1961, pp. 291-392 (text in Georgian. summary in Russian). See also Th.N. Tavadze, T.N.Sakvarelidze, Ts. Abesadze, T.A. Dvali. "On the History of Iron Manufacture in Ancient Georgia" (text in Georgian), *Restoration, Conservation Technology of Museum Exhibits* II, Tbilisi 1977 pp. 5-61.
[31] Abramishvili, op.cit., p. 300 ff.
[32] Ibid., pp. 314-316, 352, 378, cf.p. 326.
[33] Ibid., pp. 316-319.
[34] Ibid., pp. 319-322.
[35] Abramishvili, ibid., pp. 324-326; cf. p. 363.

Stratum V from the ancient settlement on Khovlegora[36] as well as an iron knife from burial No. 56 at Samtavro (Mtskheta),[37] etc.

All the materials enumerated came from Eastern Georgia. Unfortunately, there were scarcely any similar data from Colchis (Western Georgia), although Abramishvili mentioned that some objects of the Udé hoard were akin to Colchian culture.[38] He supposes that all these Early iron objects were made of metal produced locally, Proceeding from these data, he quite logically concluded that "the earliest centre of iron metallurgy is Kvemo Kartli (Lower Kartli), one of the provinces of ancient Kartli that was rich in iron ores." An iron dagger and other materials dating from the 14th-13th centuries B.C. found on the territory of Beshtasheni cemetery gave him grounds to state that this was the time of the inception and the uninterrupted development of iron metallurgy in Kvemo Kartli.[39] He is right in thinking that "the beginning of the Iron Age is the period in which uninterrupted development of iron metallurgy begins; this is possible when man has gained a knowledge of the way in which sponge iron is obtained in open hearths from iron ore extracted from deposits and has mastered the way steel weapons are forged from it."[40]

The lack of data from Colchis brought him to the logical conclusion that in Western Georgia manufacture of iron began somewhat later (about the 11th century B.C.) than in Eastern Georgia as a whole.[41] Subsequently, when new materials became available, Abramishvili came to regard the south-eastern and eastern Black Sea littoral as one of the most ancient primary centres of iron manufacture, although he considers that in the Transcaucasus "the most ancient primary centre of iron manufacture is Kvemo Kartli, a region rich in ore, as the iron objects unearthed there date from the 14th and 13th centuries B.C."[42] Moreover, he agrees with the authors who consider Anatolia to have been the primary centre of iron and steel production, where this manufacture, according to him, came into being in the middle of the second millennium B.C. From this region the technologies of iron and steel production gradually spread all over the world, including the Southern Transcaucasus, in particular, Kvemo Kartli and regions westward of it (14th-13th centuries B.C.), Southern and Western Anatolia, Syria and Palestine, Iran (13th-12th centuries B.C.), Southern Europe (turn of the second and first millennia B.C.), etc.[43]

Over 20 years ago, taking into consideration the shaky argumentation in favour of the so-called Armeno-Caucasian conception of the emergence of ancient metallurgy of iron in Armenia, S. Goginyan remarked: "The supposition that as far back as the second millennium B.C. iron was known to the ancient Armenoid tribes of the Armenian highland is based upon some laconic texts in cuneiform sources and needs to be supported by material evidence."[44]

Goginyan considers as the most important evidence of the early manufacture of iron in Armenia certain materials from Teishibaini, a pre-Uratic settlement dated by A.A. Martirosyan to the end of the 13th and the 12th centuries B.C.; at Teishibaini "there were found remnants of a forge where, alongside bronze articles, iron objects were also manufactured."[45] As for other materials from various parts of Armenia mentioned in his work they were only partial reference to the manufacture of iron, but their dating is insufficiently well supported, and they are devoid of the scientific value the author ascribes to them.

Thus, the use of iron in this or that form on Armenian territory in the 13th and 12th centuries B.C. is a quite comprehensible phenomenon, but the production of iron still has to be proved. In the light of the above, we regard as premature that "as far back as the 13th century B.C., there was, on the territory of Armenia, a sufficiently well developed (for its time) material and technological base for the manufacture of iron. Previously, all the processes of smelting and treatment of iron, prior to making the necessary objects, were done on the same site; but at the end of the 13th century B.C., these two crafts (smelting and treating iron), i.e. separate stages in the treatment of it, were basically separated. Smelting and obtaining bloom were done, as before, near outcrops of ore deposits to the surface, while smelting, apart from the preceding processes, was centered in large settlements."[46] We have, so far, no irrefutable data to support this hypothesis.

According to B.A. Shramko, on the territory of this country two most probable areas can be singled out "where, independently from each other, processes of primary mastery of iron and smelting took place." One of these was, as he supposed, in the part of Eastern Europe that was rich in deposits of iron ore, and the second - in the Transcaucasus. The latter was a secondary centre of manufacture that emerged under the influence of the realities of Nearer Asia. This author admits the possibility of such a centre existing in the North Caucasus in pre-Scythian times and agrees with scientists who consider

[36] Ibid., p. 327; cf. p. 363; L.D.Muskhelishvili. *Archaeological Material from the Khovelgora Settlement*. Tbilisi. 1978, p32, 67-71 (text in Georgian).

[37] Ibid., pp. 328-330.

[38] Ibid., pp. 325. 349; refer. 11, pp. 350-351.

[39] Abramishvili. Ibid., p. 377.

[40] Ibid., pp. 351-378.

[41] Ibid., p. 351.

[42] Abramishvili. "The Iron Age." *Georgian Soviet Encyclopaedia* 8, p. 985, 409-410; ibid., references (in Georgian), Cf. R.M. Abramishvili, T.K. Mikeladze. "On the History of Mastering Iron in the Transcaucasus and Asia Minor" *Actes du VII Congrès International des Sciences Préhistoriques et Protohistoriques*. I, Prague 1970, pp. 29-31. In specialist literature there occurs the unfounded opinion that Abramishvili "made an obvious attempt at an artificial, far-fetched fusion of the Armeno-Caucasian and Hittite theories of the origin of siderurgy", as Areshyan wrote in the summary of his Master's thesis: see G.E. Areshyan. *Iron in Ancient Western Asia*. Leningrad, 1975, p. 11 (details below).

[43] Ibid., p. 409.

[44] S. Goginyan. *On the History of Ancient Iron Metallurgy in Armenia*. 1964, No. 3, p. 229.

[45] Ibid., p. 230.

[46] Ibid., p. 234.

that, as regards the mastery of iron, the North Caucasus was in close contact with the Transcaucasus. On the other hand, Shramko categorically denies the influence of the Caucasus on the development of iron metallurgy on the territory of the European part of the USSR.[47]

Shramko explains the secondary role of the Caucasian (Transcaucasian) iron manufacture by such a "weighty" argument as the abundance of non-ferrous metals in that area. According to him, that is why the Caucasus turned from using bronze to using iron, while early iron objects (14th-13th centuries B.C.) discovered in Eastern Georgia must have been brought from Iranian Azerbaijan or North Anatolia.[48] In this connection, V.G. Kotovich remarked, with good reason, that Nearer Asia - this recognized place of origin of iron metallurgy - had a more ancient and better developed non-ferrous metallurgy than the Caucasus. This, however, did not prevent some other regions of the Near East from tackling iron smelting at a very early date.[49]

In 1971, O.A. Danielyan voiced the opinion that on Azerbaijanian territory the first iron objects appeared no earlier than the 11th century B.C.[50] Such an opinion had been voiced much earlier by other specialists studying the history of material culture of ancient Caucasian Albania.[51]

A.I. Terenozhkin strove to refute the "archaeological myth" of the profound antiquity of the beginning of the Iron Age in the Transcaucasus and even believed that iron had been imported from the North Caucasus whose inhabitants, on their part, had learnt how to smelt iron from the Cimmerians, who dwelt in the steppes farther north.[52] There are more than enough similar "conceptions" about the primary sources of iron smelting, if we may call them that. And one may stand perplexed by the fact that after a certain part of Colchian materials had been published, many researchers who claim that their theory concerning the place of origin of siderurgy is right, choose to ignore Colchis - so far the only concrete region which has the right to claim the role of the primary centre of origin of iron metallurgy.[53]

In considering the well-known letter of the Hittite king Hattusili III to King Salmanasar I of Assyria, C. Zaccagnini categorically denied the existence of the much-discussed Hittite monopoly of iron manufacture and trade.[54] Nevertheless, the thesis of the essence of the Hittite monopoly of the iron manufacture and trade is still accepted (see below).

Investigating East Georgian materials of the Late Bronze and Early Iron Ages, K.N. Pitskhelauri remarked that he had no knowledge of facts proving the existence of iron smelting workshops of the above period in that region. To support his deduction, he pointed out that all the iron smelting furnaces unearthed in Kartli turned out to be of later date.[55] In 1973, T. Wertime voiced the opinion that between 1800 and 1200 B.C. iron gradually became the leading metal in Anatolia. Subsequently, while pursuing his studies of pyrotechnology in the past, Wertime came to the conclusion that iron metallurgy originated within bronze (copper) metallurgy; but the later progress of iron manufacture was conditioned by the emergence of a critical phase in the development of society when trade in copper and other metals in regions of the East Mediterranean seaboard was checked owing to the migration of tribes. Tribes of the Black Sea littoral acquainted with the manufacture of, and trade in, iron were involved in the general movement of ethnic masses. This author thinks it possible that the spread of knowledge of the technology of iron manufacture was furthered by the fact that fuel resources (charcoal and other kinds of fuel) were exhausted in the traditional regions of metal extraction. Wertime agrees with the authors who consider that iron manufacture originated in the southern Black Sea littoral, between Samsun and Trabzon where there is an abundance of rich magnetite sands. According to him (with reference to Muhly), this region bordering on Hittite country provided it with iron.

Wertime is inclined to think it was the East Mediterranean seaboard that played the decisive role in the general dispersion of iron smelting technology, but the first systematic discovery of iron technology occurred exactly where Greek authors had pointed out - in the northern area of Anatolia and Paphlagonia (on the territory of modern Turkey). This region, which was part of the proto-Hittite state, provided the Assyrian colony Kanesh with iron. As Wertime says, the beaches of the Turkish Black Sea coast are covered with a layer of self-fluxing ferromagnetic sand with an 80 percent content of

[47] B.A. Shramko. "The Appearance and Mastery of Iron in Eastern Europe." In: *From the History of the CPSU Struggle for the Building of Socialism and Creating Communist Society in the USSR*. Kharkov, 1965, pp. 222-224. Cf. B.N. Grakov. "The Oldest Finds of Iron Objects In the European part of the USSR", *Soviet Archaeology* 1958, No. 4, p. 9; M.N. Stokova. "The Emergence of Iron and the First Skills in Obtaining it." *Proceedings of the Institute of History of the Natural Sciences and Technology*, 33, 1960, p. 233.

[48] Shramko. *The Emergence of Iron....*, p. 222.

[49] V.G. Kotovich. *Problems of the Cultural, Historic and Economic Development of the Population of Ancient Daghestan*. Moscow, 1982, pp. 167-168.

[50] O.A. Daneliyan. "On the Appearance of Iron in Azerbaijan." *Papers of the Academy of Sciences of the Azerbaijan* SSR, vol. 27/8, 1971, p. 83.

[51] G.M. Aslanov, R.M. Vaidov, G.I. Ione. *Ancient Mingechaur*. Baku,1959, pp. 120-121.

[52] A.I. Terenozhkin. "The Cimmerians and the Caucasus." *All-Union Scientific Session on the Results of Archaeological and Ethnological Research in 1970. Summaries of Papers*. Tbilisi, 1971, p. 83; Ibid., "On the History of Investigation of the Pre-Scythian period." *Scythian Antiquities*. Kiev, 1973, p. 18; ibid., "Cimmerians." Kiev, 1976, pp. 20, 200 ff.

[53] D.A. Khakhutaishvili. *Newly Discovered Monuments of Ancient Colchian Iron Metallurgy*. 151, 1977, p. 29-33.

[54] C. Zaccagnini. "KBo I, 14 e la 'monopolia' hittita del ferro." *Rivista degli Studi Orientali*, 45, Roma, 1971, pp. 11-20.

[55] K.N. Pitskhelauri. *Basic Problems of the History of East Georgian Tribes in the 15th-7th centuries B.C. (according to archaeological evidence)*. Tbilisi, 1973, p. 95, note 80 (text in Georgian, short summaries in Russian and English).

magnetic iron ore. The Cypriot, South-Turkish and Iranian master iron-smelters on being acquainted with the achievements of the iron smelters from the southern littoral of the Black Sea, were able to develop further the technology of iron smelting and the making of steel. Proceeding from information given by ancient Greek sources, Wertime recognizes the leading part played by the Chalybes and Chaldeans in initiating iron manufacture; he had supposed them to be the ancestors of modern Lazes. From 1968 to 1976, the present writer made a special investigation of the region in question, experimented in concentrating magnetic sands from the south Black Sea coast, and studied the mineralogical composition of these sands, and their ability to melt in bloomery furnaces. He ascertained that the coastal area of the Black Sea littoral provided splendid conditions for initiating iron manufacture (an ancient copper metallurgy, abundant fuels, a source of ore inexhaustible for the time, etc.) which have partially survived to this day.[56]

As we shall see below, the natural conditions of the coastal part of the southern and south-eastern Black Sea littoral, particularly the region between Samsun and Trabzon, are as good as in the region of the eastern Black sea littoral: the foothills area of seaboard Colchis (present-day Western Georgia) is rich in deposits of refractory fire-clay, magnetic sands and other kinds of iron ores, timber of various sorts for charcoal burning. This region, besides, had age-old traditions of extracting and working nonferrous metals (copper, bronze, gold, etc.).

In his monograph *Research into the History of the Most Ancient Population of Colchis and of South-Eastern Black Sea Littoral (II and I millennia B.C.)* published in 1974, T.K. Mikeladze summed up the work he had done also on the history of the metal-working tribes that resided in the region of the South-Eastern Black Sea littoral.[57] Having analyzed the data of written sources and material culture of the Transcaucasus, the Near East and the Aegean world, the author came to the conclusion that the cultural and technical influence of the ancient metal-working tribes of the South-Eastern Black Sea littoral spread far beyond the area of their habitation. Proceeding from the fact that written sources from the ancient Near East mention more and more frequently various kinds of handcrafted objects of iron (steel), and seeing that there must have appeared a relevant terminology alien to the local linguistic world, Mikeladze came to the conclusion that the makers of these objects and creators of the above terminology were the Kartvelian tribes of the South-Eastern Black Sea littoral who had been producing metals since ancient times.[58] The early contacts of the population of this region with the countries of the Near East and of the Aegean world were increasingly suggested, as the author thinks, by the term "kyanos" in Mycenaean and Ancient Greek, which may be traced to one of the ethnic names of the Kartvelian (Georgian) tribes - Chans (Tubal-Cains), as well as the term "χαλυρ" which derives from the ethnic name of the Chan tribe Chalybes-Chaldeans.[59] According to the author, the tribal names Halitu, χαλύβες, χαλδαίοι, χαλτιθ, actually denote, despite different suffixation, one and the same people, while different suffixation goes back to sources in different languages.[60]

Noting the fact that the Hittites made a wide use of iron, T.K. Mikeladze, referring to E. Laroche, supposes that iron itself, as well as the Hittite term denoting this metal ("hapalki") came into the Hittite world from the South-East Black Sea littoral: it cannot be excluded that part of the territory was occupied by Hittites.[61] The tribes inhabiting this region, the principal exporters of iron in the Mitannian and Hittite kingdoms, took part, as the author surmises, in crushing the Hittite state.[62]

Mikeladze denotes the early period of iron manufacture in Anatolia by the term "Tubal-Cainic", which came to supplant the Chalybean period by the end of the eighth century B.C. He supposes that the change of periods points to a protracted struggle for hegemony between the different tribes. It culminated in the victory of the Chalybes over the Tabals, a victory that was conditioned by the growth of iron manufacture by the winning tribe.[63]

Thus, according to Mikeladze, the South-East Black Sea littoral was the primary region where manufacture of iron originated.

In 1983 J.G. McQueen remarked that there was no evidence[64] of Hittite monopoly of iron manufacture, although in the early Iron Age Anatolia continued to play an important part in the destinies of the Near East and the Aegean Basin.[65] He holds that the use of iron increased significantly only after 1200 B.C.[66]

G.E. Areshyan, investigating the particular problem of the primary regions of iron metallurgy, came to the conclusion that there existed two early mining and metallurgical centres in Western Asia: Ionian–West Asian and Armeno-Caucasian. In his opinion, the Armeno-Caucasian region embraced the Armenian highlands and the Southern Caucasus, on which bordered the Hatti and Hattina-the main purveyors of iron to

[56] T. Wertime. "Pyrotechnology: Man's First Industrial Use of Fire." *Scientific American*, 61, No. 6, 1973, p. 681; "The Pyrotechnologic Background." *The Coming of the Age of Iron*

[57] T.K. Mikeladze. *Research into the History of the Most Ancient Population of Colchis and of the South-Eastern Black Sea Littoral.* Tbilisi, 1974, pp. 114-149 (text in Georgian, summaries in Russian and English).

[58] Ibid., p. 147.

[59] Ibid., p. 123, ff.pp. 147-148.

[60] Ibid., p. 127.

[61] Ibid., p. 139-140

[62] Ibid., pp. 147-149.

[63] Ibid., pp. 148-149

[64] J.G. McQueen. *The Hittites and their Contemporaries in Asia Minor*, Moscow, 1983, pp. 49-50.

[65] Ibid., p. 50.

[66] Ibid., p. 51.

Assyria in the ninth century B.C.[67] The complete absence (perhaps overlooked?) of reliable material evidence concerning the history of early iron ore mining deprived the author of the chance to search for concrete centres (sites) where this metal was manufactured. Although he recognized that among the peoples to whom the Greeks ascribed the invention of iron smelting (Cyclopes, Dactyls, Telchines, Chalybes) only the Chalybes were an actually existing and geographically localized people; this localization, however, was allegedly of a dispersed character, which, understandably, did not help to discover the social and cultural characterization of the development of early siderurgy.[68]

Reviving the theory of Jacques de Morgan concerning the Armeno-Caucasian "birthplace" of iron manufacture, accepted in their time by G. Frankfort, V.G. Childe and others, Areshyan attempted to prove that the Hittite conception of a generally accepted later time was the least argued,[69] as it emerged as a result of several misunderstandings and was "the greatest delusion in the history of metallurgy."[70] However, it is easy to notice that this attempt failed, as Areshyan completely disregarded the rich, factual material known to scientists since the early 1970s and widely published since then in specialist literature *viz.* Hittite cuneiform texts concerning iron, which disproves many of his conclusions about iron in the Anatolia of the Hittite epoch, as well as that mentioned above.

Areshyan does not deny the role of the Chalybes in the development of early iron metallurgy, but he considers that direct information about the Chalybes refers to a later period (to the fourth century B.C.). Previously, in the 14th-13th centuries B.C., they inhabited the Northwestern part of the Armenian highlands.[71] For some reason or other, he disregards the fact that in a part of the written sources in which the land of the iron-making Chalybes is exactly localized, it says iron was mined exclusively by the Chalybes who inhabited the coastal area, as well as the northern and north-western slopes of the modern Gümüshan and Lazistan mountain ranges. Comparing Chalybia of ancient Greek and Halitu of Hittite and Urartian texts, Areshyan concludes that "the Chalybean siderurgical centre was located in the north-eastern part of the Armenian highlands, and possibly had a certain relation to western Georgia."[72] The author, evidently, lost sight of the fact that the north-western part of the so-called Armenian highlands[73] has been, since ancient times, the zone of habitation of a Georgian-speaking (Colchian) population and is denoted by a more concrete geographical concept. Consequently, the southwestern part of the Armenian highlands was, in antiquity, inhabited by Colchian (Zani) tribes who were directly, but not "possibly," related to Western Georgia.

Areshyan's works on the history of iron do not contain any direct proof of early centres of iron smelting existing on the territory of the so-called Armenian highlands. Disregarding I.A. Gzelishvili's works and those by other authors on discoveries in Colchis, and arbitrarily assigning to the Chalybes the territory of the Armenian highlands, he takes this argument as a basis to support his theory. Such an approach to the solution of the problem may be qualified as a scholarly misunderstanding: actually the iron-making Chalybes inhabited a concrete region near the Black Sea (Pontus Euxinus); since the Homeric epoch documents have attested that they were an ethnic group, separate from the Chaldes, belonging, together with the Chaldes, to the Colchian branch of the Kartvelian (Georgian) tribes.

In the 14th century A.D., the Empire of Trebizond comprised two separate regions- Chalybia (Χαλυβια) and Chaldea (Χαλδια).[74] Proceeding from this, we should see in Strabo's communication that "in antiquity, the contemporary Chaldeans were called Chalybes,"[75] a reflection of the fact that in Strabo's time the Chalybes of the Black Sea coast were dominated by the Chaldeans. Even Strabo failed to notice any ethnic difference between them. The evidence afforded by later written sources allow us to surmise that in antiquity some of the Chaldeans (Chans) of the Black Sea Coast who had mastered the technique of iron smelting had been called Chalybes ("makers of iron"). By degrees this term, originally denoting a concrete trade (handicraft) came to denote a certain group of communities and became an ethnonym. As concerns the ethnic nature of the Chaldeans of the Black Sea coast, we might point to the fact that ancient Armenian written sources use the name "Chaltik" to denote the zone occupied by the group of Kartvelian (Georgian) tribes, the Chans.[76] According to the well-known Armenian scholar A. Adonts, "Tsanika or Chaldea, the land of the Tsans or Chaldeans occupied the highland region formed by the Parkhar range, between the Chorokh river and the Black Sea coastline, as far as

[67] G.E. Areshyan. "The Most Ancient Centres of Iron Metallurgy in Western Asia and the Eastern Mediterranean Littoral." *Proceedings of Yerevan University*, 1971, No. 3 (24), pp. 124-138; by the same author: Iron in Ancient Western Asia. Summary of Master's thesis, Yerevan, 1975, p. 8 ff. These works do not clearly differentiate between metallurgy and metal forging: not all regions where iron was forged were also mining regions. Consequently, the presence of smithing is not irrefutable evidence of the existence of iron metallurgy in the given region.

[68] Areshyan. "The Most Ancient Centres...", p. 124

[69] Areshyan. *Iron in Western Asia*, pp. 8-9.

[70] Ibid., p. 10.

[71] Areshyan. "The Most Ancient Centres...", p. 124.

[72] Areshyan. "The Most Ancient Centres...", p. 135.

[73] "...part of the Armenian highlands on the territory of the USSR (the so-called Transcaucasian highlands) occupies the entire Armenian SSR, the southern part of the Georgian SSR and the western part of the Azerbaijan SSR". *Great Soviet Encyclopaedia*, vol.II, 1970, p. 249. The term "Armenian highlands" is a modern, purely geological concept, having no ethnical or political content. Therefore, when there is a possibility to point out concretely a region or a tribe that made concrete contribution to the treasury of human culture, the constant use of names of geographic regions is not quite justified

[74] Michael Panaretos. *Trebizond Chronicles* (Materials on the History of Georgia and the Caucasus, No. 33). Tbilisi 1960, pp. 31-34, 38-39, 44, 58-59, 75-76.

[75] Strabo, XII, 3, 19. *Geography*.

[76] Movses Khorenatsi. *History of Armenia* (Georgian translation from ancient Armenian). Tbilisi, 1964, p. 162.

Trebizond..."[77] "The Tsans are one of the numerous tribes inhabiting a country which in more ancient times was known as Chaldea."[78] As Movses Khorenatsi says, "Caesar Tacticus was killed by his men in Chanica (Chanivk) or Chaltica of the Pontians."[79]

The weakness of Areshyan's argumentation, as well as of all his predecessors and followers[80] lies in the fact that it lacks concrete data on the processes of iron smelting. Without such data, any logical reasoning can only be qualified as a working hypothesis. The above is well illustrated by numerous "conceptions" concerning the primary regions where iron metallurgy was initiated. Therefore, we regard as premature the author's conclusion that "in the vast West-Asian zone the inception of the iron industry was not exclusively conditioned by an ordinary technical discovery, nor by the existence of certain ethnical groups or a group of iron-makers."[81]

Dwelling on the problem of the "industrial mastering of iron", Areshyan surmises that it happened in Armenia, and "partially in Georgia" in the two last centuries of the second millennium B.C. Putting it in his words, "such a dating implies the independent emergence of iron manufacture in this region conventionally termed by us the Ponto-Caspian mining and metallurgical centre of early iron, independent of other centres of primary siderurgy in Western Asia." De Morgan's theory of the spread of primary iron metallurgy from an Armeno-Caucasian primary centre should be rejected, as well as the possibility of iron metallurgy penetrating into the Transcaucasus from the south. And all this is supposedly confirmed by the fact that "the above region had the required natural resources-abundant deposits of iron ore."[82]

In a work published in co-authorship in 1977 and devoted mainly to the eastern Black Sea littoral, R. Maddin, J. Muhly and T. Wheeler conclude that "although iron was known as a metal that lent itself to processing, if not during the whole Bronze Age, at least during its greater part"; despite this fact, about 500 iron objects dating from before the end of the second millennium B.C. are purely ornamental and are not intended for household or military needs.[83] It was only after the end of the 12th century B.C. that for reasons not wholly known to us, the use of iron grew rapidly, the sphere of its usage broadened (economy, military needs, construction work, etc.), and, naturally, the use of copper and bronze was gradually restricted. According to Snodgrass, at the beginning of the first millennium B.C. (1050-900 B.C) in one of the regions of Greece the ratio of iron and bronze objects was as follows: knives, 1 to 15; swords, 1 to 20; spearheads, 9 to 30 in favour of iron. According to the above mentioned authors, the process of making articles of bronze is simpler than their manufacture of iron; besides, damaged articles of bronze could be smelted several times, but iron was not, in those days, amenable to re-casting. These authors conjecture that future archaeological research may show the reason why the use of bronze decreased, for it had the advantage over iron smelted in a bloomery hearth in that it smelted at a lower temperature, could be cast, was less brittle, resisted corrosion, etc.[84]

But all this, as the authors think, refers to the period before the technique of cementation had been discovered. After forging, "steeled iron" becomes twice as solid as highest-grade bronze. By the beginning of the 10th century B.C., blacksmiths already knew the process of cementation of iron, although there are certain data pointing to the fact that the technique of cementation of iron was already known in the 12th century B.C.[85] The discovery of the technique of tempering (the first literary mention of it occurs in the Odyssey),[86] together with cementation and subsequently drawn to reduce brittleness, was of great significance to the manufacture of high-grade steel objects.

Judging by the general trend of the publication in question, one can surmise that the authors regard the Eastern Mediterranean seaboard as the region where techniques of manufacturing steel (cementation, tempering and drawing) were invented.

It is noteworthy that as late as in 1977, V.G. Kotovich stated erroneously that "not in a single one of the Caucasian monuments belonging to the latter half of the second millennium and the turn of the second and first millennia B.C. do iron and even steel objects occur regularly, nor are there any traces of the manufacture of iron."[87]

However, further acquaintance with materials from the Colchian region brought about a radical change of V.G. Kotovich's views.

While investigating the problem of the emergence of iron metallurgy in Daghestan, Kotovich also considered the general problem of the emergence of iron metallurgy in

---

[77] A. Adonts. *Armenia in the Time of Justinian*. Yerevan, 1971, p. 26.
[78] Ibid., p. 65
[79] Movses Khorenatsi. Op.cit., p. 162.
[80] See, e.g. S.T Khachaturyan *The Ancient Culture of Shirak*. Yerevan, 1975, pp. 232-233
[81] G.E. Areshyan, *Iron...*, p. 28
[82] G.E. Areshyan. *The Mastering of Iron in Armenia and in the Southern Caucasus*, 1974, No. 2, p. 211; by the same author: Iron..., p. 26.
[83] R. Maddin, J. Muhly and T. Wheeler. "How the Iron Age Began." *Scientific American*, 1977, No. 237/4, p. 122

[84] Maddin et al., ibid. According to these authors, the limit tensile strength of sponge iron is approximated to 40,000 lbs. per square inch, while that of bronze is 60,000 lbs. per square inch. Lengthy smithing may increase the strength of iron to 100,000 lbs. per square inch, while cold smithing of bronze- to 120,000 lbs per square inch. At the same time, bronze had another advantage- it melted at temperatures that could be attained in ordinary ancient furnaces (1200°C), while pure iron melts at 1537°C; without attaining such a temperature, iron cannot be cast.
[85] Ibid., pp. 126-127.
[86] Ibid., pp. 128-129. "The log hissed in the eye of the Cyclops, like a great axe or adze when the blacksmith immerses them in water, 'curing' them; this is the only way steel can be hardened" (*Odyssey*, IX, 390).
[87] V.G. Kotovich. "Some Questions of Ancient Copper Metallurgy in Connection with the Problem of the Emergence of Iron Metallurgy in the Caucasus." *Soviet Archaeology*, 1977, No. 3, p. 75.

the Caucasus as a whole and reviewed all the various viewpoints on the problem (the Transcaucasian conception proceeding from J. de Morgan's and B.A. Kuftin's Near-Asian conception).[88] Bringing forward concrete material, Kotovich refuted, convincingly and logically, the argumentation of the supporters of the hypothesis according to which the existence of non-ferrous metallurgy, allegedly, hampered the early spread of iron in the Caucasus. On the other hand, he remarked "none of the extant viewpoints on the origin of metallurgy in the Caucasus can be recognized at present as finally confirmed; all the more, it is not the only possible one."[89] He was quite right in stressing that the problem cannot be solved without creating the indisputable factual basis. Despite this, Kotovich ventured a number of suppositions about the ways in which iron metallurgy appeared and took root. Speaking of the achievements of Caucasian metallurgists in the third millennium B.C. thanks to the use of furnaces with induced draught, he agrees with A.A. Baikov's opinion that such furnaces afforded the opportunity of smelting not only copper ore but iron ore as well. Thus, as he rightly noted the Caucasus was technically prepared, as early as the third millennium B.C., if not earlier, to perform any metallurgical operation known in antiquity.[90] As he supposed, the earliest iron objects dating mostly from the end of the second and the beginning of the first millennium B.C. found in the Transcaucasus (Caucasus) were made from metal obtained by smelting chalcopyrite or siderite employed for preparing ochre, a mineral pigment.[91]

In the light of these and other data, Kotovich is inclined in favour of the theory of multi-centred independent regional development of iron metallurgy.[92] He concurs with those who believe that "iron metallurgy had been developed in Asia Minor and Iran during basically the same historical period as in the Transcaucasus."[93] And rejects the opinion of B.A. Shramko according to whom the early (14th-13th century B.C.) iron objects found in Georgia were imported from Iranian Azerbaijan or Northern Anatolia.[94]

Following the research of Georgian scholars, Kotovich emphasized the special role of ancient Colchian material for studying the primary centre of iron smelting technique. In the light of this, he concludes that "there is no doubt that the Transcaucasus area was one of the regions where the first steps of mastering the technique of iron smelting took place." The earliest stages of the mastery of iron metallurgy are not yet fully understood but the data enable us to characterize the later phases of the process. These facts give us grounds for concluding that the ancient metallurgists of the Transcaucasus started experimenting with the technology of iron processing and research into the durability of the metal very early—towards the middle or perhaps from the beginning of the second millennium BC. These experiments resulted in the mastery of the technology enabling the production of "raw and cemented steel" as early as the 14th-13th centuries BC. (with reference to the works of Abramishvili, Th.N. Tavadze, A.T. Ramishvili and D.A. Khakhutaishvili).[95]

Taking into consideration copious material on the Transcaucasus and the Near East, V.G. Kotovich concludes that "these facts prove the correctness of the opinion of the researchers who consider that one of the oldest centres of iron metallurgy appeared and took shape in the Transcaucasus.[96]

In a publication that came out in 1980, A. Snodgrass summed up the work he had done in investigating history of iron metallurgy in the Mediterranean Seaboard.

According to Snodgrass, written data on the early iron metallurgy were not always reliable, as the coming of iron into common use in the overwhelming majority of regions coincided with a period of decline in culture and chronicles of the time are not always complete and sufficiently reliable; as for later sources, mostly Greek ones, they are a far cry from the inception of iron metallurgy. "We do not reject the statements of Strabo and other geographers concerning the location and condition of some Mediterranean centres of metal production in the late Classical Period, but theauthenticity of their existence is to a great degree indirect."[97]

As regards Europe, Snodgrass singled out three stages in succession in the use of iron, dating from various times and in different regions, the first characterized by the use of iron for the manufacture of ornaments or objects not used in household chores. In the second stage, the production of iron is increased, but its use to cover economic and military needs is restricted as compared

[88] V.G. Kotovich. *Problems of the Cultural, Historical and Ecomomic Development of the Population of Ancient Daghestan.* Moscow, 1982, p. 165 ff.

[89] Ibid., pp. 168, 182.

[90] Ibid., p. 175.

[91] Ibid., pp. 181-183. Unfortunately, the author's argumentation in support of this view is not always convincing; besides, he does not clearly differentiate metalworking from metal ore mining (v. p. 182 where the house of the smelter blacksmith is mentioned as proof of the existence of mining), although he is very well aware that "The ancient metallurgy of iron, like copper-bronze metallurgy, was divided into mining and metalworking." We consider unfounded the author's suggestion that iron obtaining consisted of mining and dressing the ore, and getting fuel, while metal working, i.e. the making of iron consisted of obtaining iron by smelting in a bloomery furnace (p. 183)

[92] Kotovich, *Problems* ... pp. 183, 188-189.

[93] Ibid., p. 192.

[94] B.A. Shramko, "The Appearance and Mastery of Iron in Eastern Europe", p. 222 (see above)

[95] V.G. Kotovich. Op.cit., pp. 193, 194, 211-212.

[96] Ibid., pp. 211-121. D.A. Khakhutaishvili, "On the Primary Centres of the Inception and Development of Iron Metallurgy." *Jubilee collection devoted to the 100th anniversary of I.A. Djavakhishvili.* Tbilisi, 1976, pp. 97-103 (in Georgian); D.A. Khakhutaishvili. "A Contribution of the Kartvelian Tribes to the Mastery of Iron Metallurgy in the Ancient Near East." *Acta Antiqua Academiae Scientiarum Hungaricae*, 22, fasc. 1-4, 1974, p,337-348.

[97] A.M. Snodgrass. "Iron and Early Metallurgy in the Mediterranean." *The Coming of the Age of Iron*, New Haven, 1980, p. 335.

with bronze. At the third stage iron prevails over bronze as a metal in practical use, although the production of bronze objects continues. According to him, during the first stage the technique of cementing iron was still unknown.[98] In his opinion, Urartu is a remarkable example of the stability of bronze industry, while other regions had developed a wide use of iron for economic needs (Cyprus, the Aegean region, the Levantine littoral, etc.).[99] As he supposes, in the East Mediterranean littoral Cyprus is an earlier centre where the economic use of iron became common at the end of the ancient period of Cyprian culture III B, which corresponds to 1050 B.C.[100] The technology of metal-working spread from Cyprus throughout the ancient world, both in the West and in the East.[101] It should be noted that Snodgrass, as well as the authors he cites, mention only iron objects as proof of this statement, without indicating any example proving the existence of local smelting of iron. If Cyprus and the Aegean world were actually the primary centres of iron metallurgy, it should be explained why the ancient Greeks ascribed the discovery of secrets of iron production to other regions and other peoples.

Proceeding from this conception, Snodgrass does not, naturally, share the widely accepted viewpoint about the Hittite (Anatolian) centre of origin of iron, as based upon not very reliable primary sources. Not having a very clear idea of the data drawn from Hittite cuneiform texts concerning iron, he cites only the well-known letter of Hattusili III. In his opinion, this letter, dated approximately to the 13th century B.C., says nothing about a Hittite monopoly of iron production, but points to the absence of it. Besides, the early appearance of iron (Snodgrass' first stage, when iron was not used for the needs of economy) does not indicate that the region in question soon moved on to a wide use of iron.[102] He claims that the most advanced iron producing regions of Anatolia (e.g. Cilicia) were about 200 years behind in their development as compared with the East Mediterranean Centres, while "Urartu is still more retarded in the mastery of the new metal, although it is situated near to the land of the Chalybes to whom legend ascribes the discovery of iron."[103] As will be shown below, Snodgrass was not very well informed about the state of things in the Colchian Chalybean problem .

Considering the problem of the development of the "Iron Age" in Egypt, Snodgrass thinks it a unique phenomenon that Egypt, having a knowledge of iron since very early times and possessing a highly developed economy, passed from bronze industry to iron remarkably slowly. Noting that Egypt gives the most convincing proof of the fact that it lacked any historical need of a transition from bronze to iron, he leaves open the problem of the cause of such a phenomenon[104] without noticing that the natural conditions of economic activities in the lower reaches of tropical rivers (the Nile, the Tigris, the Euphrates, the Yangtse and Hwan Ho, etc.) did not favour the rapid spread of more effective instruments of labour than those of bronze.

After a study of the cuneiform texts known to her and having to do with early iron, Jane C. Waldbaum remarked that only one Hittite text dating from the 13th century B.C. speaks of the process of smelting iron; that is the above-mentioned letter of the Hittite King Hattusili III. All the other texts mention the iron traded or the use of this metal for various purposes.[105]

J. Waldbaum does not support the idea of a Hittite nopoly of iron production and refers to the paper by J. Muhly published in the same collection as hers.[106] Dwelling upon all the known archaeological facts of the use of iron (including meteoric iron) beginning with the epoch of Early Bronze, and taking laboratory analysis into consideration, she accepts Snodgrass' view that the Iron Age took shape when this metal became less costly and was finally recognized as surpassing other metals for manufacturing instruments of labour and weapons of war. In her opinion, this moment came about in the 10th century B.C. in Mesopotamia, and somewhat later in Europe and in other, more easterly countries and regions. The preceding period which began about 1200 B.C. in the East Mediterranean littoral may be regarded as a period of transition from bronze to iron and termed "The Early Iron Age." She thinks "there is no doubt the Hittites used iron for both weapons of war and for agricultural implements and applied a better technology of iron production than their neighbours did. If the Hittites did have a certain secret, they did not yet regard the iron made in this way as suitable for economic needs."[107]

As she says, "we have very little reliable material, if any, on the use of smelted metal in the Late Bronze epoch." Referring to the well-known letter of Hattusili III to Shalmaneser I, she remarks that in many texts more is said about the use of iron and trade in it than about its manufacture in the Late Bronze Period."

It should be noted that the author of the above publication was not well acquainted with the data of cuneiform, including Hittite sources and had scarcely any knowledge of Transcaucasian materials and, naturally, of any relevant literature which accounts for some of her unfortunate fallacies and misconceptions. This also explains her statement that "we know very little about the techniques of making iron at the early stage of the Bronze Age and the period of transition...; we know that smelted iron existed as early as the third millennium (maybe

[98] Ibid., pp. 336-337.
[99] Ibid., pp. 338-341.
[100] Ibid., pp. 341-344, 355-356.
[101] Ibid., pp. 344-345, 355.
[102] Ibid., p. 357.
[103] Ibid., pp. 358.
[104] Ibid., pp. 364-365.
[105] J.C. Waldbaum. "The First Archaeological Appearance of Iron and the Transition to the Iron Age." *The Coming of the Age of Iron*, New Haven, 1980.
[106] Ibid., p. 81
[107] Ibid., p. 81-82 ff.

earlier), although it is not known whether it was produced by chance or intentionally." Unfortunately, Waldbaum overlooked not only Russian publications, but also specialist literature in the English language.

Investigating the problem of the Iron Age taking root in Western Iran, V. Pigott turns to Caucasian materials, including those from Transcaucasia. Pigott voices the opinion that according to archaeological data obtained during the last few years, the transition from bronze to iron took place at the end of the second and in the early first millennium B.C.[108] He considers more convincing the opinion of D. Young who thinks that the radical changes that occurred in Western Iran in the second half of the second millennium B.C. (new techniques in pottery production, extra-mural cemeteries, the erection of fortresses, etc.), including the coming of iron metallurgy, were connected with the arrival of Iranian tribes from Eastern Iran. On the other hand, Pigott cites the opinion of C. Burke who holds that the tribes that had brought these novelties into Western Iran had come from the north.[109]

Summing up the result of recent investigations into the history of early iron in Western Iran (Iranian Azerbaijan), Pigott suggests a more exact periodization of the Early Iron Age: Iron I – 1450/1350 – 1100; Iron II 1000 – 800 B.C.; Iron III 700 – 550 B.C.[110] There is very scanty material, as he says, to substantiate the existence of the conjectural Iron I period (an iron ring from the grave of Hasanlu V, some not very exactly dated material from Jeoya - iron slag and some smelted haematite ore) which cannot even give an answer to the question as to the nature of the metal-smelting process. Not far from the region in question is cemetery "A" of Tepe Sialka. In one of the graves of this necropolis there was an iron dagger and an iron punch.[111]

The Iron Age in Western Iran is characterized by a growing presence of iron goods in nearly every context and also by the appearance of regional cultural peculiarities instead of overall cultural uniformity. It was at the height of cultural changes that the real Iron Age came, as Pigott supposes, when iron became more important than bronze and covered new spheres of economic life.[112]

Touching upon the situation in the neighbouring Caucasus, Pigott notes that in central Caucasus the early stage of iron production is now dated to the 11th or 10th centuries B.C. According to him, the more evidence is collected in the region of the Caucasus, the less likely it becomes that the production of iron on a greater scale comparable with the amount of implements and weapons from Talysh began long before the 10th century B.C.[113] Pigott takes into account the necessity of co-ordinating Transcaucasian and Iranian materials; nor does he deny the possibility of Iran being influenced from "centres of metal production situated far West."[114] He holds that materials from Hasanlu discovered by R. Dyson point to the period encompassing the 10th and the 9th centuries B.C. when ironsmiths found iron to be malleable and a useful metal capable of serving various ends. The technique of iron production remained primitive but its use for various purposes grew considerably.[115]

Pigott shares R. Playner's opinion that Urartu was "the key to the Early Iron Age in Iran and the Caucasus, that the influence of Iran spread over all the neighbouring regions, bringing with it iron, especially weapons and the technique of their production. On the other hand, he was well aware of the fact that Urartian epigraphy mentions iron only once in the tine of Sarduri II (8th century B.C.), while iron objects from this region are not much older than the date just mentioned.[116] Dwelling upon iron goods from the Sailka II graves, Pigott admits the possibility that part of those objects appeared as a result of well-established trade relations with a certain producing centre in the West.[117] He agrees with van den Berg's opinion that from the 14th to the 11th centuries B.C. bronze was the only utilitarian metal in Western Luristan; from the 10th to the 8th century B.C. bronze was still used for making weapons and tools, while iron was used for making ornamental bracelets, rings, pins, anklets, bridles, etc., and, now and then, ceremonial weapons. Bimetallic objects also occur (hairpins, bracelets, daggers, etc.).[118]

In conclusion Pigott remarks that so far no archaeological object directly connected with smelting iron has been found in Western Iran.[119]

Proceeding from a work published jpintly by Cook and Aschenberg, Pigott states that iron was obtained by chance: the master-smelters of the Pre-Iron Age used iron ore (haematites or magnetites) as flux when smelting copper, thus creating the potential chance of obtaining iron instead of copper; little by little man discovered the properties of iron, thus preparing the pre-requisites for the beginning of the Iron Age.[120]

On our part, we must note here again that as regards Western Iran, there are no reliable materials to prove that at least part of the iron used in the Early Iron Age was smelted locally. In their joint publication "Metallurgy and Ancient Man", T.S. Wheeler and R. Maddin, following R.F. Tylecote, came to the conclusion that the technique of carbonising iron was discovered after the year 1200

[108] V. Pigott. "The Iron Age in Western Iran." *The Coming of the Age of Iron*. New Haven, 1980, p. 418.
[109] Ibid., p. 419.
[110] Ibid., p. 420.
[111] Ibid., p. 421.
[112] Ibid., pp. 422-423
[113] Ibid., pp. 422-424
[114] Ibid., p. 430
[115] Ibid., p. 432.
[116] Ibid., p. 437.
[117] Ibid., p. 438.
[118] Ibid., p. 447.
[119] Ibid., p. 450
[120] Ibid., p. 452

B.C. in one of the coastal regions of the East Mediterranean,[121] while "blacksmiths resorted widely to carbonisation of iron at least since 1000 B.C." According to these authors, "the iron knife studied by Tholander (found in Idalion) and dated to the 12th century B.C., had been carbonized and, possibly, even tempered."[122] Transcaucasian materials are completely lacking in the above work too. R.F. Tylecote who had made a special study of ancient furnaces, crucibles and slags came to the conclusion that in antiquity "in Asia Minor and in Europe smelted iron was obtained sometimes by chance." As he says, remnants of early iron production are very scanty in Asia Minor and in Persia. In a special work he could mention only two rather doubtful cases when remnants of early iron manufacture were found. One is F. Dorner's find in Arsameia, near Nymphaios, close to Yenikalé, of the remnants of a pre-feudal cup-shaped iron-smelting furnace, 100 cm in height, 40 cm in diameter and with a wall 3-4 cm thick.[123] The second case is a large dump of iron slag found by H.G. Bachmann near Malatya, in Sirzah. He thinks these slags may be dated to the 8th century B.C., as they were produced by the blooming process and are very much like the slags from ordinary African furnaces.

When reading Tylecote's work one feels somewhat vexed by the paucity of his knowledge about the history of metallurgy in Transcaucasia.

In the opinion of the authors of *The History of the Ancient World*, which came off the press in 1983, in the second millennium B.C., from the Middle East to Europe inclusively, the tribes inhabiting North and West Asia Minor held a monopoly on iron production. The Hittite kings, like their predecessors, the rulers of Proto-Hittite city-States, kept this monopoly strictly guarded, as it was one of the sources of replenishing the state treasury. After the fall of the Hittite state, the royal monopoly on iron ceased, but as late as the first millennium, "iron deposits were kept secret by the tribes living in their vicinity, a tribe the Greeks called Chalybes, (presumably a western proto-Georgian tribe), but then there was no one to prohibit the export of iron, and it was freely exported through the upper valley of the Euphrates and the city-states of the North-Syrian confederation to the south (from the ninth century B.C) and northwards, through Ionian colonies to the Black Sea littoral and farther west (from the eighth-seventh centuries B.C.)." According to the authors of the above-mentioned book, most of the campaigns waged by Assyria, Urartu, Phrygia and Media were attempts to seize "the iron way" and gain possession of this new strategic raw material.[124] The authors conjecture that in some regions of Syria, the Transcaucasus and Europe the secret of iron production was revealed soon after the fall of the Hittite state.[125]

Thus, in the work under study coming from the pen of authoritative Orientalists of this country, the widely accepted opinion on the monopoly of Hittite rulers on technological secrets of iron smelting is also accepted. On the other hand, the role of the Chalybes in the development of iron metallurgy is recognized, while the possibility is not excluded that they (Chalybes) belonged to the Kartvelian ethnic world. It seems to us that this is an attempt to reconcile or even unite two hypotheses concerning the inception of iron metallurgy - the Hittite and Chalybean hypotheses.

Analyzing the state of affairs concerning the research into the history of ancient iron, Ivanov came to the conclusion that the researchers who consider Asia Minor to be the "homeland" of terrestrial iron are quite right. As he puts it, the opinions of earlier investigations should be checked, and it should be recognized that the inventors of the technique of producing iron and steel, i.e. "good iron", were not Hittite, but proto-Hittites, from whom the former inherited their knowledge; it is not excluded that Hittite kings established a monopoly on iron trade, thereby promoting the retardation of a fast and wide dispersion of this metal.[126] Ivanov assigns the first appearance of steel objects to the 13th century B.C. (an iron axe from the lower part of Khattusa), and considers as contemporaries the first evidences of steel production and its tempering in the Aegean world (on Cyprus, a region for which the last Hittite kings fought and where the earliest evidences of iron date back to the 13th-12th centuries B.C.). In his opinion, "...the Greeks who had by this period already gained a knowledge of the technique of iron and steel making used in Asia Minor, for a long time preserved a memory not only of the Oriental origin of this technique, but also of the people – Chalybes (Χαλυβες), "iron-makers" whose name itself speaks of the nature of their metallurgical discoveries and is reflected in the toponyms that survived up to the Byzantine period, and, possibly later, in contemporary Asia Minor.[127]

In connection with the problem of the Chalybes, Ivanov cites N. Marr's opinion expressed in a letter back in 1915 that linguistic links show that the Khatti were a Caucasian people identifiable with the Chalybes of ancient Greek authors,[128] and states that Marr's foresight is corroborated

[121] Tamara S. Wheeler and Robert Maddin. "Metallurgy and Ancient Man." *The Coming of the Age of Iron*. New Haven, 1990, p. 115.

[122] Ibid., p. 121.

[123] F.K. Dörner et al. "Arsamaia at Hymphais." *Jahrbuch des Deutschen Archaeologischen Instituts*, Berlin 80 (1965), 88-235. R.F. Tylecote. "Furnaces, Crucibles and Slags." *The Coming of the Age of Iron*. New Haven, 1980, pp. 211-212.

[124] I.M. Dyakonov, V.A. Yakobson, N.B. Yankovskaya. "Common Features of the Second Period of Ancient History." *History of the Ancient World*, II.Moscow, 1983, pp. 5-6. Cf. Carlo Zaccagnini, "Kbo I 14 e la monopolia hittita del ferre." *Rivista degli Studi Orientali*. 45,1971, pp. 11-20, where Hittite monopoly for iron production and trade is categorically denied.

[125] Ibid., p. 6.

[126] V.V. Ivanov. *The History of Slavic and Balkan Names of Metals*. Moscow, 1983, pp. 95-96, 107-108.

[127] Ivanov. *The History of Slavic and Balkan Names of Metals. Moscow*, 1983, p. 129. with reference to X Plantho. "De Geographia Pontica II. Les Khalibes: nom de people ou qualification professionelle". *Journal Asiatique*, 202, 1963, Nos. 2-3, pp. 298-309.

[128] N.Y. Marr. *On the Language and Writing of the Abkhazians*. Moscow, 1938.

by the latest archaeological and linguistic data.[129] He expresses the opinion that "the very name of the Chalybes and the kind of steel they manufactured is a transformation of the ancient name of iron which in its original Khattish form (xaflki) contained a consonant cluster impossible in Greek f/wlk, transformed to –λυβ. The information from ancient sources concerning the Chalybes' place of habitation on the shores of the Black Sea conforms to the conclusion about the role of the northern littoral of Asia Minor in an earlier period, lately arrived at by scholars in the light of the newly-found ancient Hittite texts, in particular, in the story of Tsalpa.[130] Not only the Greek name of the Chalybes and of the steel they produced is a transformation of the Khatti xaflk, but also the general name for metal (and of the basic metal of the Copper and Bronze Ages- copper): Mycen. Gk. ka-ko, Gk. Hom. χαλκος (in particular, meaning "weapon of metal", Mycen. kake-u "blacksmith" (χαλκευς)." According to Ivanov, "...and a number of other ancient Greek terms connected with other mythological notions about metals, particularly about iron, may be phonetic variants of the same Khatti term with regular dissimilation of aspirates, according to Grassman's term. This may explain not only the Colchae and the Colchis, but also the denomination of the inventors of forging, the Telchinae Τελχ, where the labialised kw may reflect Khatti labialization of -f-, cf. υ-υβ in χαλυβ."[131]

Thus, in the work mentioned above, the Chalybes-Khatti are described as the discoverers of iron smelting, and a real basis has been revealed for recognizing the South-Eastern Black Sea littoral as the original "home" of iron metallurgy. However, there is no identity between Chalybes and Khatti, as Ivanov thinks, if the thesis is correct that "Khattish belongs to the North-Caucasian languages, having especial affinities with the North-Western (Abkhazian-Adyghé) languages." ("History.... Metals", p. 133). The Chalybes had no kinship with North-Caucasian tribes. As has long been recognized in scientific literature, the Chalybes were one of the West-Kartvelian tribes.

A noteworthy publication on the problem of the production and use of iron in Asia Minor in the second millennium B.C. is a comprehensive paper by Jana Siegelova, a Czech Hittitologist who collected, commented and translated into German most of the principal data of Hittite texts concerning iron by the beginning of the 1980s.[132] According to these data, iron was used in Anatolia ever since the time of the Ancient Hittite kingdom (the pre-Hittite period, i.e. the period of the "Cappadocian" tablets when processing of the metal in question was already known in Asia Minor is not dwelt on in her paper), iron was used as a precious metal for making objects of symbolic significance, objects of luxury, etc. In the periods that followed, mostly in the epoch of the New Kingdom of the Hittites (14th-13th centuries B.C.) iron was more and more widely used, which is manifested not only by a growing number of iron objects, but also by a greater variety of types. Besides objects of ritual use, iron was used in manufacturing a great number of knives, daggers, swords, battle weapons and other objects. As the author of the paper supposes, Hittite texts say nothing about the technique of iron production; however, proceeding from the data offered by Cappadocian tablets and the letter of King Hattusili III (v.supra), iron was smelted in the land of the Hittites. The sources of raw material were plentiful, and iron-makers were the earliest to stand out in a separate group from other "makers" of metal.

In his new publication "The Production and Use of Iron In Central Anatolia According to Hittite Cuneiform Texts" and proceeding from the newest materials and an original interpretation of some cuneiform texts, as well as archaeological materials, G.G. Giorgadze demonstrates that in the time of the New Kingdom of the Hittites, especially at the close of the period of existence of the Hittite state, iron was no longer used for producing objects of luxury,[133] but was employed for making ornaments, idols, cultic objects having a symbolic significance during rituals and religious feasts, etc.; over and above that, it was used for manufacturing objects indispensable in everyday life. After analysing relevant cuneiform texts, Giorgadze remarks that various kinds of iron are mentioned in them: "iron", "celestial iron", "black iron", "black celestial iron", "good (ritual) pure iron", "hearth iron."[134] He considers that "black iron" was not always meteoric, as most researchers think, but denoted a certain kind of this metal obtained by smelting, but by some properties as yet unknown to us (strength, colour) resembling meteoritic iron. The author's conjecture is supported by the fact that, according to the cuneiform texts, many different things were made from "black iron" (long vessels, daggers and knives, the bases of statuettes, etc.). At the same time, "black iron" could not have been a metal of low quality, as, according to one cuneiform text (KUB XLII, 78 1113) it was used along with precious stones and gold, although it was of a lower quality than "good gold" of which a small amount was produced.[135] The author supposes that the Hittite Ē babalki in the ritual of Ishtar (KUB XXIX, 73, 12) denoted a smithy and is another evidence of the Hittites' early practice of forging iron.[136]

[129] Ivanov. Ibid., p. 107.
[130] Ivanov, ibid., p. 98, with reference to M.J. Mellink "Ancient Metals trade." *Science*, 185, 1974, pp. 52-53. Mellink "Archaeology in Asia Minor," *AJA*, 42/2, 1974, No. 2.
[131] Ibid., p. 98
[132] J. Siegolova. "Gewinnung und Verarbeiten von Eisen in Hethistischen Reich im 2 Jahrtauzend v.u.z." *Annals of the Naprstek Museum* 12, Prague, 1984, pp. 71-168.
[133] G.G. Giorgadze. "The Production and Use of Iron in Central Anatolia, According to Hittite Cuneiform Texts". In: *The Ancient East. Ethnic and Cultural Ties*. Moscow, 1988, pp. 238-261.
[134] Ibid.
[135] Ibid., p. 245
[136] Ibid.,

Considering the well-known letter of Hittite King Hattusisi III,[137] the only text containing a direct mention of the production of iron, Giorgadze remarks that "the request of the Assyrian king was evidently about sending him "good iron", not merely an object made from it. The fact that the Assyrian king was interested in just this kind of iron that only the king of the Hittites possessed. Hattusisi promised to send him some "good iron" as soon as it was manufactured. All the above makes indisputable the priority of the Hittites over the other peoples of the Near East as regards the production of "good iron." The author writes that he disagrees with the opinion of J.S. Waldbaum who calls in question the Hittite priority in the mastering of the production of iron[138] (v.supra).

On the basis of data in cuneiform texts and archaeological material, Giorgadze comes to the conclusion that in the 14th-13th centuries B.C. a transition took place in Anatolia from the manufacture and use of precious and ritually sacred iron to using iron for making a broader assortment of iron ware. This is a very important factor for determining the beginning of the early period of the Iron Age in Central Anatolia in the 14th-13th centuries B.C., after which iron came to be widely used in Anatolia, and in ancient Egypt generally from the 12th century B.C. to modern times.[139]

Touching upon the geography of the production of early iron, Giorgadze remarks that Hittite texts do not provide any information concerning the sites where the raw material was mined or iron-smelting furnaces and other components of iron manufacture were located. He regrets that " the archaeology of Asia Minor is also silent on this point." He has "no doubts, however, that in Anatolia of the Hittite epoch all this existed", seeing that Eastern Asia Minor, in particular the mountains of Pontus and Taurus were rich in iron ore in ancient times, while the Northern and North-Eastern regions of Anatolia, i.e. the later Pontus, produced iron not only in the epoch of the Hittite state, but much earlier- when the Khatti, the inventors of smelting iron—hapalki—from ore were in existence. Giorgadze surmises that the production of iron on the above territory went on even after the fall of the Hittite kingdom, which is testified to by the fact that since the 9th century B.C. the Assyrians had been receiving great amounts of iron as tribute from districts of Hatte, i.e. regions to the West of the Euphrates in the Eastern part of Asia Minor. He shares the opinion of Ivanov that subsequently the Greeks applied the name of Chalybes to the iron-making tribes dwelling on the Pontic coast just because of their manufacturing skills; he conjectures, however, that, ethnically speaking, these tribes belonged to the West-Georgian group of Kartvelian (Georgian) tribes.[140]

Summing up his research, Giorgadze concludes that "Hittite cuneiform inscriptions quite unambiguously point to the existence of a well-developed Hittite iron metallurgy, in particular in the period of the New Kingdom (14th- 13th centuries B.C.) when large amounts of iron implements were manufactured in Khatti (both from meteoric and terrestrial iron) which would have been unthinkable without a sufficient mastery of iron production and iron metallurgy. There also was a well-developed technique of manufacturing a variety of iron objects of various weights and sizes, meant not only for ornamental or cultic, but also for utilitarian, purposes. At times, "iron objects prevailed over those of bronze and copper…"[141]

Hittite cuneiform texts prove without any doubt that iron was widely in use in the Hittite Kingdom, although they mention the production of this metal within the limits of Hittite territory; however, Hittite sources say nothing at all about mining iron ore. Having no doubt whatever that iron ore was mined and iron was produced on Hittite territory, we do not exclude the possibility that a certain amount of iron ore and iron was imported into Central Asia Minor from regions bordering on Hittite lands, first and foremost from South-Western Georgia, particularly from Colchis. We trust that the present publication of materials from Colchis will provide a definite answer to the problem in question.

Traces of the production of ancient iron on the territory of Western Georgia were discovered, as has been mentioned, soon after the end of WWII when archaeological research was renewed.

In Guria (Ozurgeti district, Georgian SSR), on the territory of the Anaseuli and by the source of Natanebi River, in a locality named Korisbudé, O.M. Djaparidze registered over 50 "centres." The excavation of two such "centres" gave him the possibility to conjecture that they belonged to iron-smelting workshops: a lump of slag taken for a laboratory analysis was found to contain 80.3 percent iron.[142] Excavation of these sites did not yield any datable material and the problem of the chronology of this group of monuments remained unsolved.

Several years later, N.V. Khoshtaria published a work in which she summed up what she had done in an archaeological investigation of the lower reaches of the rivers Supsa and Natanebi. The material unearthed corroborated the presence of large-scale production of iron and determined the source of raw material

[137] Ibid. "As for the 'good iron' you have written to me about, there is none in Kitsuvatne, in my House of the Seal. The iron for making it is bad. I have written to them that they should start making good iron. They have not as yet finished. As soon as they have done it I shall send it to you. At present I am sending you a dagger blade of iron."

[138] Ibid.

[139] Ibid. and footnote referring to A. Goetze *Kleinasien*, Munchen, 1957, p. 186.

[140] Ibid.

[141] Ibid.

[142] O.M. Djaparidze. "Exploratory Expedition in Guria." *Proceedings of the S.N. Djanashia State Museum of Georgia*, 16/B, 1950.p. 111.

(magnetitic sand).[143] According to Khoshtaria, "remnants of iron-smelting furnaces occur on almost the whole territory of Anaseuli, as well as on that of Ekadia village, towards the Natanebi railway station. In all, there are over 50 points where are clusters of such remnants: near the present-day cemetery of Anaseuli, near a bamboo grove, on the road to Bzvani, near the Orapo river, on the slope above the confluence of the Natanebi river (1 km. from the Natanebi railway station). One of the sites of the greatest concentration of iron forge remnants is called "Navrudja" or "Narudja."[144] Unfortunately, Khoshtaria had no materials to enable her to determine the date of these monuments of iron production.

In 1953, in the middle reaches of the Choloki river we discovered another group of monuments of iron production. This group was viewed, in 1959, by I.A. Gzelishvili who already had a certain experience of archaeological research of monuments of iron production. This marked the beginning (in 1960) of a systematic study of ancient iron metallurgy in Colchis by a joint archaeological expedition from the I.A. Djavakhishvili Institute of History, Academy of Sciences of the Georgian SSR (at present the I.A Djavakhishvili Institute of History, Archaeology and Ethnology of the Academy of Sciences of the Georgian SSR) and the Batumi Research Institute, also of the above Academy (at present the A.N. Berdzenishvili Research Institute, Batumi, of the Georgian Academy of Sciences).[145]

After the untimely demise of I.A. Gzelishvili, in the archaeological woek in which he was engaged was held up for years; in 1970 it was resumed by the present writer. As a result of our investigations, we succeeded in ascertaining that there was, in pre-Antique Colchis, an iron-smelting and manufacturing area, quite large for its time, comprising about 400 sites grouped into four principal centres in the foothill zone of Ancient Colchis (Eastern Black Sea littoral).

A quarter of a century has passed since the initiation of the study of the above group of monuments. Interesting and highly important materials for science have been collected, preliminary publications of which were issued after a certain group of sites had been unearthed.

[143] N.V. Khoshtaria. "Archaeological Investigations at Ureki." *Materials of the Archaeology of Georgia and the Caucasus*, Tbilisi, I, 1955, pp. 71-73.

[144] Ibid., p. 72

[145] I.A. Gzelishvili. *Main Results of Excavations of Iron-Smelting Workshops Found in Adjaria in 1960-1961*, pp. 31-32; idem *Iron Smelting in Ancient Georgia*, pp. 5-6.

CHAPTER I

THE CHOROKHI METALLURGICAL SITE

The Chorokhi manufacturing site of the Colchian centre of ancient iron metallurgy was discovered in 1960 by the joint archaeological expedition of the N.A. Berdzenishvili Research Institute of the Georgian Academy of Sciences and the I.A. Djavakhishvili Institute of History, Archaeology and Ethnology of the Academy of Sciences of the Georgian SSR.

The site is situated on the left bank in the lower reaches of the Chorokhi (the Acampsis of the ancient authors), on the territory of the Tskhinvali state farm, the villages of Avgia and Akhalsopeli, in the mountain zone of the Gonia citrus fruit state farm. The objects of investigation are dispersed over the hills and slopes of the foothill zone, near the convergence of spurs of the Meskheti and Lazistan mountain regions (Fig. 1).

In the foothills not far from the seashore, the soil is characteristic for the subtropic zone of the Eastern Black sea littoral: yellow tufogenic clays with an admixture of magnetite grains. This feature assigns them to the refractory group of fire-clay, which, along with certain other conditions, is one of the bases for the emergence and development of ancient bloomery metallurgy.

Sandy beaches and all the New Black Sea terrace are rich in magnetite, one of the prerequisites for the initiation of bloomery furnace metallurgy.

Thus, deposits of fire-clays and refractory clays, an abundance of various fuel and magnetitic sands created the basis for initiating the ancient bloomery technique of iron production. As will be shown below, more ancient traditions of metal making also promoted it. The opinion is not unfounded that the Chorokhi basin was one of the principal centres of ancient non-ferrous metallurgy in the Near East, while the South-West littoral of Georgia is regarded as the region where the universally known Colchian Late Bronze culture took shape.

Excavations of the Chorokhi production centre were initiated by I.A. Gzelishvili in 1961, but were unavoidably discontinued in the same year. It was only in 1979 that work was resumed.

1. THE IRON-SMELTING WORKSHOPS "CHARNALI 1"

The iron-smelting workshops "Charnali 1,"[1] like the whole of this manufacturing site, were found in 1960, but the excavations of the latter with its two iron-smelting workshops were carried out in the autumn of 1961.[2]

The Charnali 1 iron-smelting workshops are situated in the same area, on the territory of the former Charnali (at present Tkhinvali) state farm, Khelvachauri district, Adjarian ASSR, in the gully of the Putkris-ğelé brook, a left-bank confluent of the Charnalistqali River which falls into the Chorokhi from the left. The archaeological site was situated on a promontory-like hill overgrown with wild subtropical bush. On the right and left the hill is bordered by deep gullies, on the left opening into a gorge. The territory around the site to occupied by a tea plantation and various other subtropical plants (Fig. lA).

The soil forming the territory where the remnants of iron-smelting work are located is composed of tufogenic formations from the Tertiary period, above which are strata of yellow diluvial clays two or more metres thick depending on the level of a concrete stratum.

An area of 225 m^2 (15 × 15mi.) was excavated. The depth of the archaeological stratum was about 1 metre from the contemporary surface. This revealed: remnants of two smelting furnaces, the ruins of stone tables (anvils) for primary forging with wooden beetles of the spongy mass extracted from the surface and dumps of waste.[3]

I.A. Gzelishvili regarded the finds from the site as contemporary and dated them to the 8th-7th centuries B.C.[4]

However, further investigation demonstrated that these iron-smelting furnaces belonged to an earlier time, and were of different ages: archaeomagnetic analysis demonstrated that smelting furnace "Charnali I, 1" is about 100 years older than smelting furnace "Charnali I, 2."[5] An attentive scrutiny of the same data brings us to the same conclusion: in the lining of furnace "Charnali I, 2" a round cake of slag was found which most probably belonged to an earlier furnace, i.e. "Charnali I, 1." Incidentally, such "cakes" were found everywhere when waste dumps from ancient iron manufacture were excavated.[6]

[1] I.A. Gzelishvli, *Iron Smelting in Ancient Georgia*, pp. 47-52, Figs.14 A, 14 B, 15 A-B, pl. II, 4, IX, 6, 7, 13.

[2] The personnel of the expedition: I.A. Gzelishvili (head), D.A. Khakhutaishvili (deputy head); members: A.K. Inaishvlili (director of the Batumi Research Institute), A.T. Ramishvili, L.B. Chkhaidze, V.M. Poplavskaya, V.K. Mitaishvili, R.K. Atoyan.

[3] I.A. Gzelishvili. *Iron smelting in Ancient Georgia*, pp. 47-48, figs. 14A, 14B, 15; pl.II, 4, IX,B, 6, 7, 13.

[4] Ibid., p. 51.

[5] Z.A. Chelidze. "Results of Archaeological Analysis of Some Archaeological Objects". In: *Problems of Ancient History. Caucasian-Near-Eastern Collection*, V, Tbilisi 1977, pp. 142-145.

[6] Gzelishvili. Op.cit., p. 49.

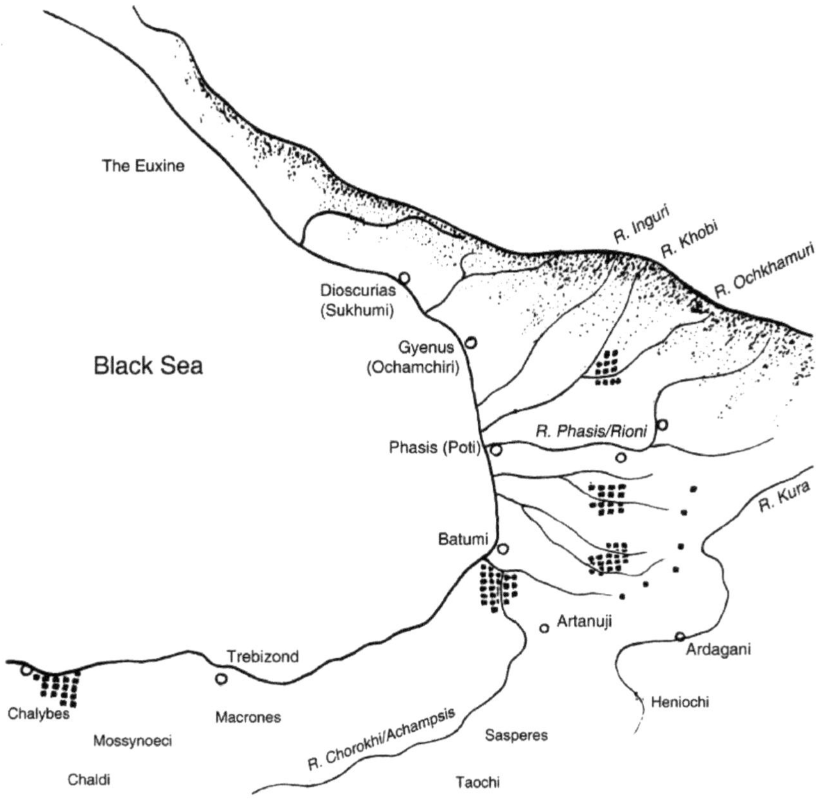

Fig. 1. Location of Iron-Manufacturing Centres in Ancient Colchis.

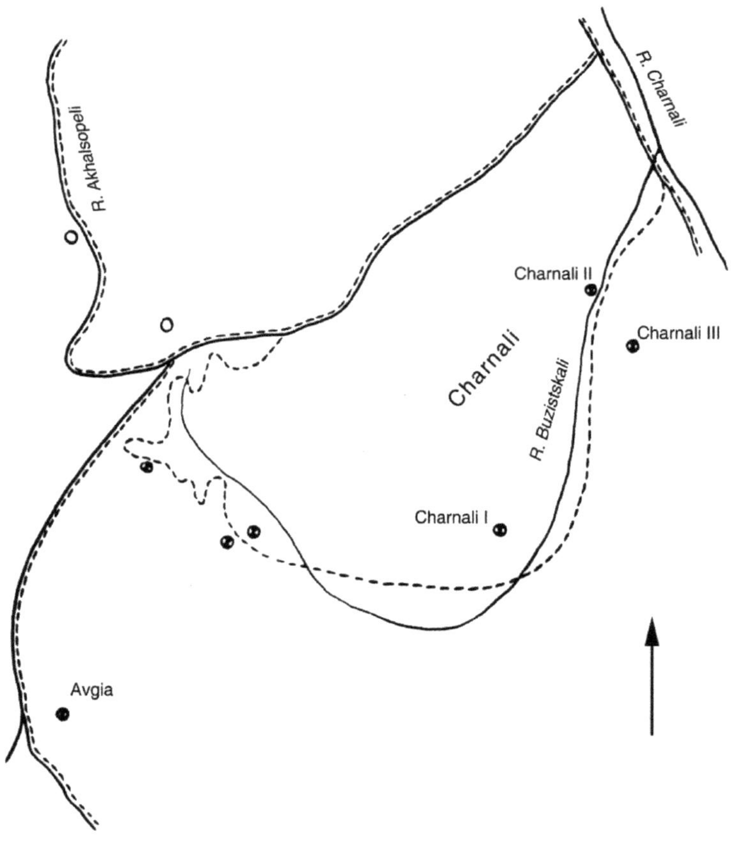

Fig. 1A. Location of Separate Groups of Iron-Smelting Workshops of the Choroki Production Centre.

In addition, some vessel fragments found in waste dumps appear at the turn of the second and the first millennia B.C. This variety of vessels is of the same kind as thick-walled burnished Colchian ceramic vessels, having on the outside broad fluting cut into the surface[7]

a. Iron-Smelting Workshop "Charnali I, 1"

This workshop has come down to us consisting of the smelting furnace, the table (anvil) for working the bloom and the waste dump (Fig. 2).

The remnants of the workshop were buried under a layer of soil and yellow clay up to 100 cm thick. The area was over-grown with various kinds of shrubs.

The remnants of the smelting furnace were unearthed in an area near the south part of the promontory in the centre of an open space.

The surviving part of the furnace is at the bottom of the pit shaped like an upturned truncated pyramid. The inner space of the furnace was filled in: the upper part by a layer of soil, under which was yellow clay with admixtures of bits of iron slag, clay lining of the furnace, stones fallen from its masonry, fragments of nozzles of bellows and pottery sherds.

The upper part of the furnace was lined with stones (two rows), while the lower ending in a spherical bowl lined with a thick layer of fire-clay. All data suggests that the whole inner surface of the furnace was lined before being filled with charge.

The furnace was 120 cm deep, 80 cm wide at the top, 50 cm in the middle, 10-15 cm at the bottom.

The table (anvil) for working the bloom survived as fragments of stone pavement with surface 300×150 cm, under which the clay was burnt red by a high temperature. Next to the pavement was charcoal storage for charcoal paved with sherds and enclosed by a low border.

The waste dump was not tar from the smelting furnace on a sloping area exposed to the south. This area was about 60 m.sq., the thickness of the cultural layer attaining 100 cm The dump contained: a great many lumps of slag, (up to 6 cubic metres) with a "thick" greyish patina characteristic of pre-antique slags, fragments of clay furnace lining, also greyish (ash-coloured) and reddish fragments of bellow nozzles. Digging the dump revealed several sherds of ceramic vessels, including some of a large vessel ornamented with broad cannelures done by "cutting" (Figs. 16, 9)[8] and a fragment of the handle of a pot dark-grey in colour.

The workshop is dated to the end of the second millennium and the early first millennium B.C.,[9] which is suggested by the pottery,[10] as well as by archaeomagnetic data, according to which the last smelting in the "Charnali I, 1" furnace was done about the middle of the 11th century B.C. (declination = –28.0°, inclination = 60.0°, tension = 0.492 oersted)."[11]

The other workshops in the same area turned out to be of somewhat later date.

b. Iron Smelting-Furnace "Charnali I, 2"

The iron-smelting furnace "Charnali I, 2" (Fig. 2) came down to us consisting of the smelting furnace itself, the table (anvil) for forging the bloom and the waste dump.

The iron-smelting furnace "Charnali I, 2" is beside the furnace "Charnali I, 1", but 100 cm from it (Fig. 2). It is in a pit fashioned like an upturned cone-shaped pyramid. The upper part is lined with rough stone, the lower part has a thick layer of fire-clay. It should be mentioned that a round "cake" of iron slag, 35-40 cm was used in the upper part of the pit lining. The inner space was filled in with clay mixed with ashes, scorched stones, bits of slag, fragments of clay coating of the furnace and bits of nozzles. At the bottom of the pit was a large charcoal layer in sizable pieces 'smeared' with clay.

The depth of the pit was 130 cm; at the top it was 90cm wide, in the middle 60 cm, at the bottom 10-15 cm.

The table (anvil) survived as a flat area of clay burnt red on which were fragments of the clay-plastering of the furnace. Most probably, the owners of this workshop had put to use the remains of an old one.

It was found that the waste dumps had mingled, over time, with the waste of the more ancient workshop. The dump contained a great amount of iron slag (over $3m^3$), fragments of the clay coating of the furnaces, of nozzles, pieces of charcoal, ashes and stones from the furnace facing and the stone table. Most stones are reddish in clefts, the result of being subjected to high temperatures.

A store of fire-clay was found at the rim of the dump, under a layer of ashes.

According to archaeomagnetic data obtained by Z.A. Chelidze, the last smelting in this workshop was done around the middle of the 10th century B.C. (declination = -14°, inclination = 64.5°, tension = 0.368 oersted).[12] Such a dating is not contradicted by the archaeological data

[7] Ibid., pl.IX, 6. D.A. Khakhutaishvili. "On the Chronology of the Colchian-Chalybean Centre of Ancient Iron Metallurgy". In: *Problems of Ancient History. Caucasian-Near-Eastern Collection*, V, 1977, p. 136. ff.

[8] Gzelishvili. *Iron Smelting in Ancient Georgia*, pl. X.

[9] Gzelishvili assigned this object to the 8th-7th centuries B.C.

[10] See: Khakhutaishvili, *On the Chronology of the Colchian-Chalybean Centre of Ancient Iron Metallurgy*, pp. 119-141.

[11] Z.A. Chelidze. *Results of Archaeomagnetic Analysis of Some Archaeological Objects*, p. 145. Unfortunately, charcoal for radiocarbon analysis was not taken in 1960-1961.

[12] Z.A. Chelidze. *op.cit.*, p. 145.

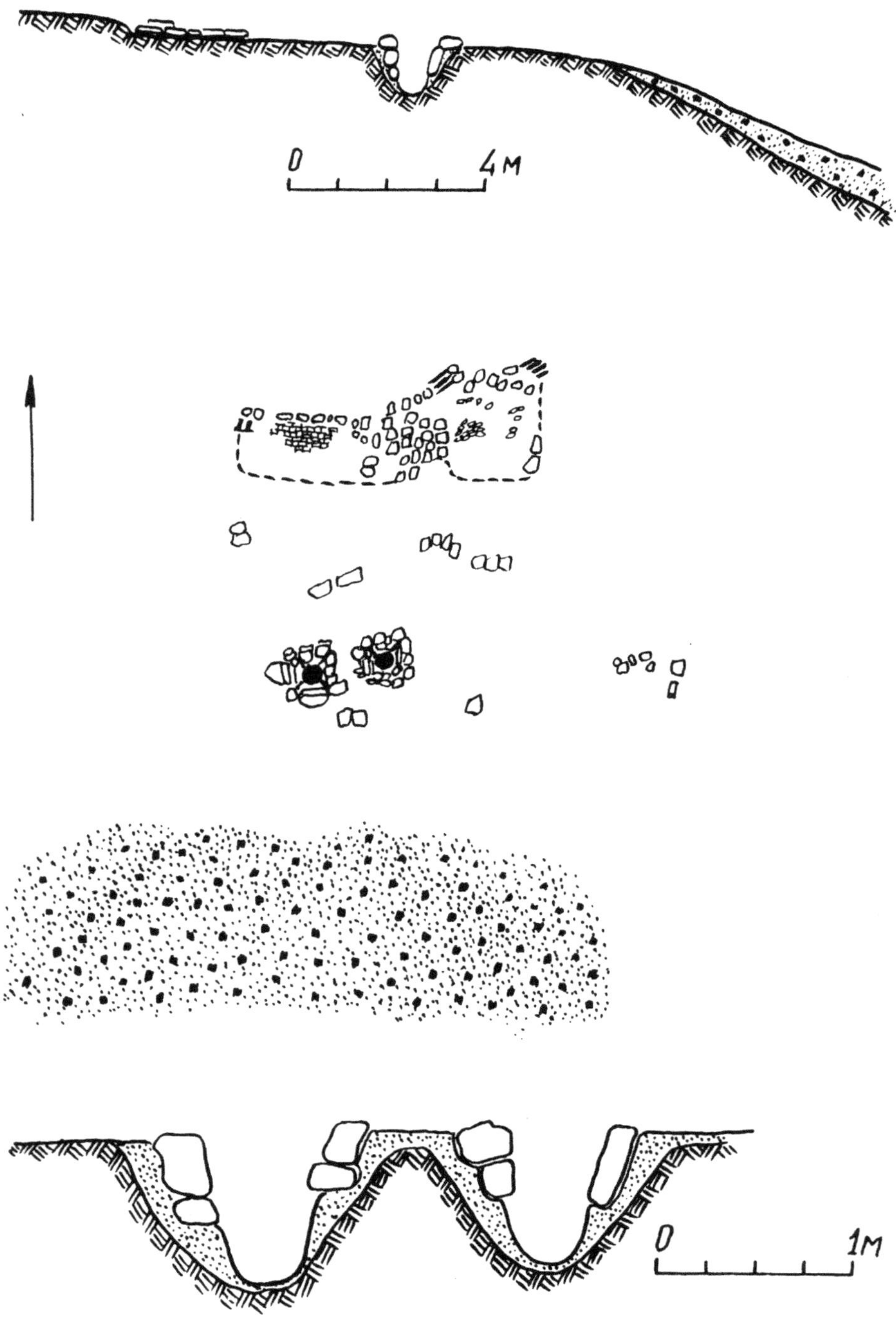

Fig. 2. Iron-Smelting Workshops "Charnali I." Ground Plan and Sections.

which, although scanty, are not devoid of a certain meaning.

As for the structure of the iron-smelting furnaces belonging to the above workshops, we may note that despite their comparatively simple structure they were sufficiently efficacious. Judging by the fact that within furnaces whose stone lining has survived intact some of the stones have been affected by rather high temperatures, we may suggest that the above-ground parts of the furnaces which have not come down to us were built of rough stone plastered with clay. It may be surmised that the upper part in question was taken to pieces after every smelting and rebuilt after being charged again.

2. THE IRON-SMELTING WORKSHOPS "CHARNALI II"

The archaeological site of ancient iron manufacture which we have termed "Charnali II"[13] was first noted in 1961 and excavated in 1979 by the archaeological

[13] A preliminary report by D.A. Khakhutaishvili: "Main Results of Work Done by the Archaeological Expedition in South-Western Georgia in 1979." In: *Monuments of South-Western Georgia* XI, Tbilisi, 1982, pp. 10-14, figs.4, 4.

expedition of the N.A. Berdzenishvili Research CeInstitute of the Georgian Academy of Sciences.[14]

The monument lies on the left bank of the Putkris-ğelé brook, 1 km. downstream from workshop "Charnali I", on the state farm's tea plantation. This site was chosen because when a road was being laid across the plantation a bulldozer had damaged the waste dump and, it appeared that the furnace itself might have been impaired.

Excavation revealed that in this place, almost on the very bank of the brook, under a layer of soil and yellow clays, there were the remnants of two iron-smelting workshops of pre-antique times (Fig. 3). It also showed the timeliness of the work done by the expedition for the investigation and conservation of the monument.

Laterite soils (the result of the decay of various tufogenic Tertiary formations in a subtropical climate) are widespread in this area; above them are deposits of yellow diluvial clays.

The first traces of ancient iron manufacture were unearthed at a depth of 25 to 30 cm.

a. Iron-Smelting Workshop "Charnali II, 1"

The iron-smelting workshop "Charnali II, 1" was found to consist of the smelting furnace, the composite stone table (anvil) for the hot forging of the bloom, and the waste dump thrown down into the brook in 1975 when the road was being laid.

The iron-smelting furnace is arranged in a pit shaped like an overturned truncated pyramid (Fig. 3). The upper part of the pit is lined with stones of various origin, while its lower part is a hemispherical cavity thickly plastered with fire-clay. All the inner surface of the furnace was plastered in the some way: acted upon by the intense heat, stones of the facing were tinted red or mauve, and the clay plastering was cinder-coloured. At a depth of 15-20 cm the ground around the furnace was reddish tinted.

The inner space of the furnace was filled in with yellowish clay mixed with bits of slag, fragments of bellow nozzles and of an earthenware vessel adorned with broad fluting, ashes and roughly ground charcoal. At the bottom of the furnace there was also a layer of charcoal with an admixture of clay. Samples of the charcoal to be subjected to radiocarbon analysis were handed over to the dating laboratory of Tbilisi State University (sample No. TB-286).

The depth of the surviving part of the furnace was 140 cm, width at the top 110 cm, in the middle 90 cm, and at the bottom 50 cm (Fig. 3).

The composite table (anvil) was well preserved. It was placed to the north of the furnace, at a distance of 150 cm, its surface being 200×190 cm. The pavement is of rough basalt and river boulders. East and west of it are two flat areas paved with clay lining of the furnace. These areas are bordered by a low brim of one row of small stones. Of the west area only some parts remain, while the East one is in a good state of preservation (Fig. 3).
As mentioned above, the dump belonging to this workshop was thrown down into the brook in 1975 when the road was built, and buried under a lot of earth. We managed to find only a few bits of slag and plastering of the furnace.

To date the workshop, we had at our disposal the data of the analysts of charcoal from the bottom of the furnace. The Dating Laboratory of Tbilisi State University determined the age of the sample as 2670±50 years, which corresponds to 720 B.C.[15] According to the table of P.E. Daymon and his colleagues, the age of this sample is 2750 ± 102 years, corresponding to 800 B.C.[16]

Thus, the "Charnali II, 1" iron-smelting workshop functioned in the 8th century B.C., according to radiocarbon analysis. This dating is not contradicted by fragments of a ceramic vessel ornamented with broad fluting done by the "cutting" technique.

b. The Iron-Smelting Workshop "Charnali II, 2"

This workshop was located on the same site as workshop "Charnali II, 1." As usual, it consisted of remnants of the furnace, the composite stone table (anvil) and the waste dump.

The iron-smelting furnace "Charnali II, 2" is adjacent to the iron-smelting furnace "Charnali II, 1", 80 cm away (Fig. 3). The lower hemispherical cavity is plastered with a thick layer of fire-clay, while the top part was lined with cleft basalt and river boulders. The soil around the furnace had turned reddish from great heat.

Within, the furnace was filled with yellow clay mixed with pieces of slag, stones from the revetment, fragments of bellow nozzles and clay plastering, fine-ground charcoal and unidentified pottery sherds. At the bottom of the iron-smelting furnace there was a layer of roughly ground charcoal samples which were sent to the laboratory in Tbilisi University.

The surviving part of the furnace has a depth of 150 cm, upper width 90 cm, in the middle 60 cm, at the bottom 15-20 cm.

[14] Personnel of the expedition: D.A. Khakhutaishvili (head), S.A. Gogitidze, A.M. Djavelidze, A.B. Gogua, G.N. Bibileishvili, M.A. Zarkua.

[15] A.A. Burchuladze, G.I. Togonidze. *Radioactive Dating*, IV, Tbilisi, 1985, sample TB-286.

[16] Ch.A. Arslanov. "On Corrections of Radiocarbon Dating." *Geochemistry*, 1978, No.8. p. 1161.

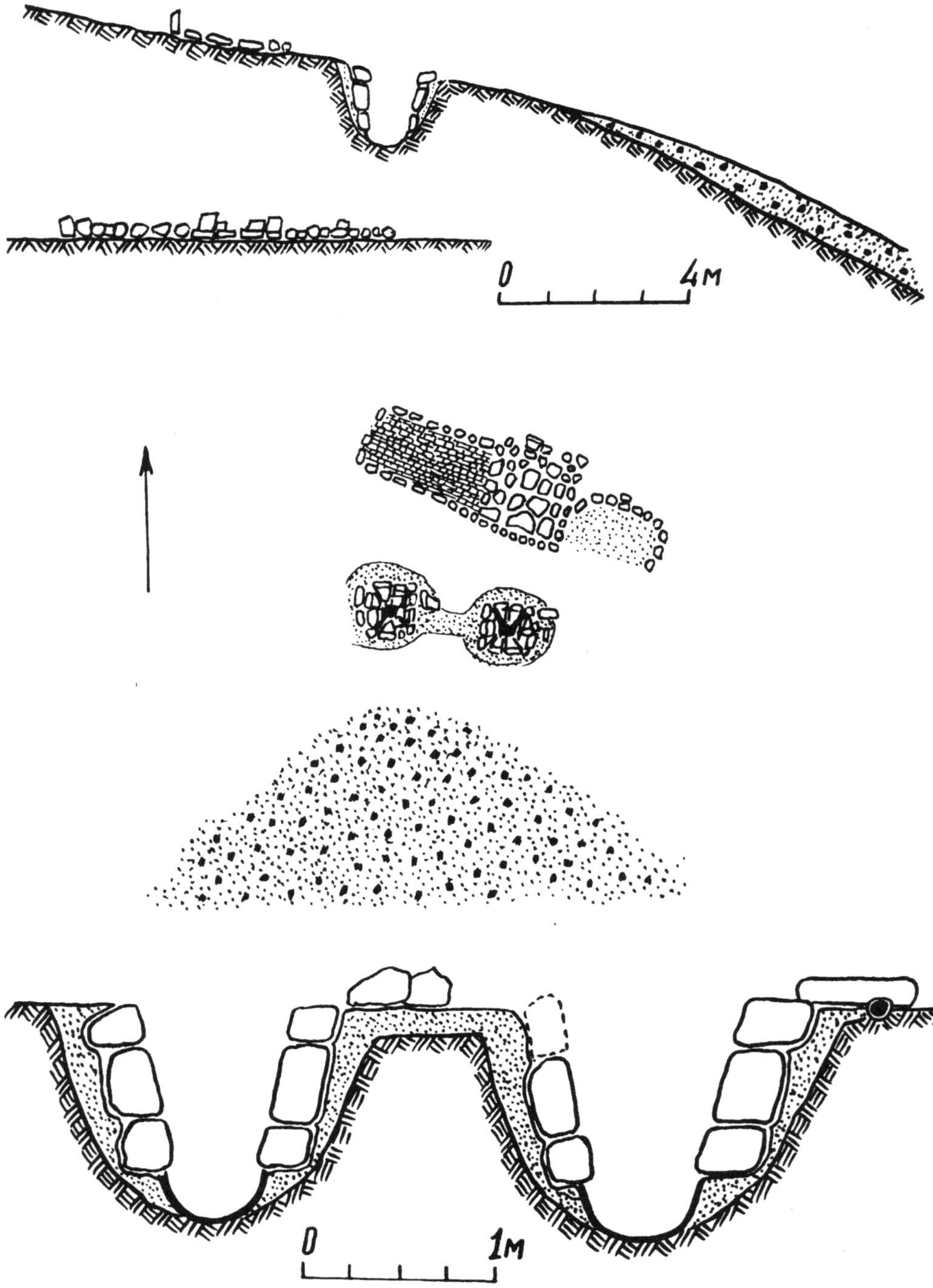

Fig. 3. Iron-Smelting Workshops "Charnali II." Ground Plan and Sections.

The owner of this smelting furnace used the same composite stone table for working the bloom as did the owner of the other workshop from this site.

As mentioned above, the waste dump had been destroyed when the road was being built. Only some of the slag and fragments of clay plastering were recovered.

According to the data determined by the dating Laboratory of Tbilisi State University, the sample taken from the bottom of "Charnali II, 2" furnace (TB-287) is dated to 2720± 102, i.e. 770 B.C.[17] The table compiled by P.E. Daymon and his colleagues determined the date of this sample at 2811±102 years, i.e. 861 B.C.

Thus, the iron-smelting workshop "Charnali II, 2" functioned several decades earlier than "Charnali II, 1."

Consequently, this part of the Chorokhi centre had been in use twice – first, in the 9th century B.C. and then in the eighth century B.C. The chronological break between

[17] A.A. Burchuladze, G.I. Togonidze. *op.cit.*, TB-287.

these workshops is also corroborated by archaeomagnetic data.

3. THE IRON-SMELTING WORKSHOPS "CHARNALI III"

The archaeological site where the "Charnali III"[18] group of iron-smelting furnaces was discovered lies on the right bank of the Putkris-ğelé brook, 200 metres away and on a mountain terrace overhanging "Charnali II." On this sloping terrace with its northern exposition tea and other subtropical plants are grown. On the northern side, the terrace touches a steep slope, on the east a comparatively gentle declivity. The terrace slopes slightly eastwards, a smooth strip 10 to 15 metres in breadth, on the east slope of which there survive a few basalt "cannon balls" of palaeovolcanic origin. The entire excavated area is approximately 575 m^2 (25 × 23 m.). On the northern and eastern slopes of the terrace there were two waste dumps.

The tea shrubs and the top layer of cultivated soil being removed, the outlines of two round red stains were unearthed at a depth of 25-30 cm evident proof of the existence of smelting furnaces.

a. Iron-Smelting Workshop "Charnali III, 1"

Of this workshop, the iron-smelting furnace, traces of the composite table (anvil) and of the waste dump have survived (Fig. 4).

The iron-smelting furnace "Charnali III, 1"[19] has come down to us as a pit shaped like an overturned truncated pyramid, two thirds of its upper part being lined with rough-hewn basalt slabs. The lower part of the furnace ending in a hemispherical cavity and indeed all the inner surfaces were plastered with a heavy layer of fire-clay. All the inner surfaces were plastered with it. The fired clay became dark-grey, ash-coloured under the effect of the intense heat, while the surface of the facing stones was purplish. Here and there the walls of the furnace had melted. The inner space was filled with clay in which scorched stones were found most probably the ruins of the subterranean part of the furnace; also found were pieces of iron slag, fragments of bellow nozzles, and of the clay plastering of the furnace. At the bottom was a layer of charcoal of which samples were taken for radiocarbon analysis.

The dimensions of the furnace were: height 120 cm; width in the upper part 90 cm, in the middle 65 cm, at the bottom 15-20 cm.

The table (anvil) remained as a scorched flat area (200×100 cm) and several scattered scorched stones.

As we have already mentioned the waste dumps were on the east and north slopes of the terrace, but as the workshops were built on the same plot, it is hard to tell which dump belonged to which. It should be noted that the dump on the east slope was almost destroyed when the ground was cultivated. Some fragments of iron slag picked up from this dump are covered with a thick grey patina characteristic of slags of the period. According to the data given by the Dating Laboratory of Tbilisi University, the examples of charcoal from the bottom of iron-smelting furnace "Charnali III, 1" belong to the turn of the ninth and the eighth centuries B.C. (2750± 50 = 800 B.C.). According to the table presented by P.E. Daymon and his collaborators, these samples are assigned to 890 B.C. (2840±102 = 890 B.C.).[20]

b. Iron-Smelting Workshops "Charnali III, 2"

The surviving parts of this workshop are: the iron-smelting furnace, traces of the composite table (anvil) and the waste dump. Right there, between two iron-smelting furnaces, a hemispherical pot (height 30 cm, diameter 80 cm) was dug (Fig. 4). This is the first time such an element of a smelting workshop has been registered.

The "Charnali III, 2"[21] iron-smelting furnace is in the shape of an upturned truncated pyramid, the upper part of which is lined with roughly-hewn slabs of basalt, the lower part plastered with a thick layer of fire-clay. The intense heat gave the plastering of the furnace a dark-grey, ashen tint, while the stones, especially the ones at the top, turned purple. The furnace walls, particularly the upper part faced with stones, are melted in places.

Within the furnace filled in with clay there were pieces of slag and charcoal, fragments of the clay plastering and of the nozzles of bellows. At the bottom of the furnace the layer of roughly-ground charcoal was badly smeared with clay. The earth around the furnace along both the vertical and horizontal perimeter had turned red at depth of 15-20 cm from the intense heat.

The iron-smelting furnace is 125 cm high/deep, 100 cm, wide at the top, 90 cm in the middle and 25-30 cm at the bottom.

The waste dump was on the north slope of a stepped terrace covered with thick grass. It contained a great many lumps of slag and bits of charcoal, fragments of clay plastering, bellow nozzles and ashes. Unfortunately, the dump was not intact, as archaeological material slipping down the steep incline into the brook was swept away down its course.

With regards to the date of this workshop, we can only say that it belongs to the pre-antique period. This is attested, in the first place, by the thick greyish patina on the iron slag and by the structure of the iron-smelting

[18] For the preliminary report; See: D.A. Khakhutaishvili. "Main Results..." pp. 14-17.

[19] The furnace had been conserved.

[20] Preserved *in situ*.

[21] Preserved *in situ*.

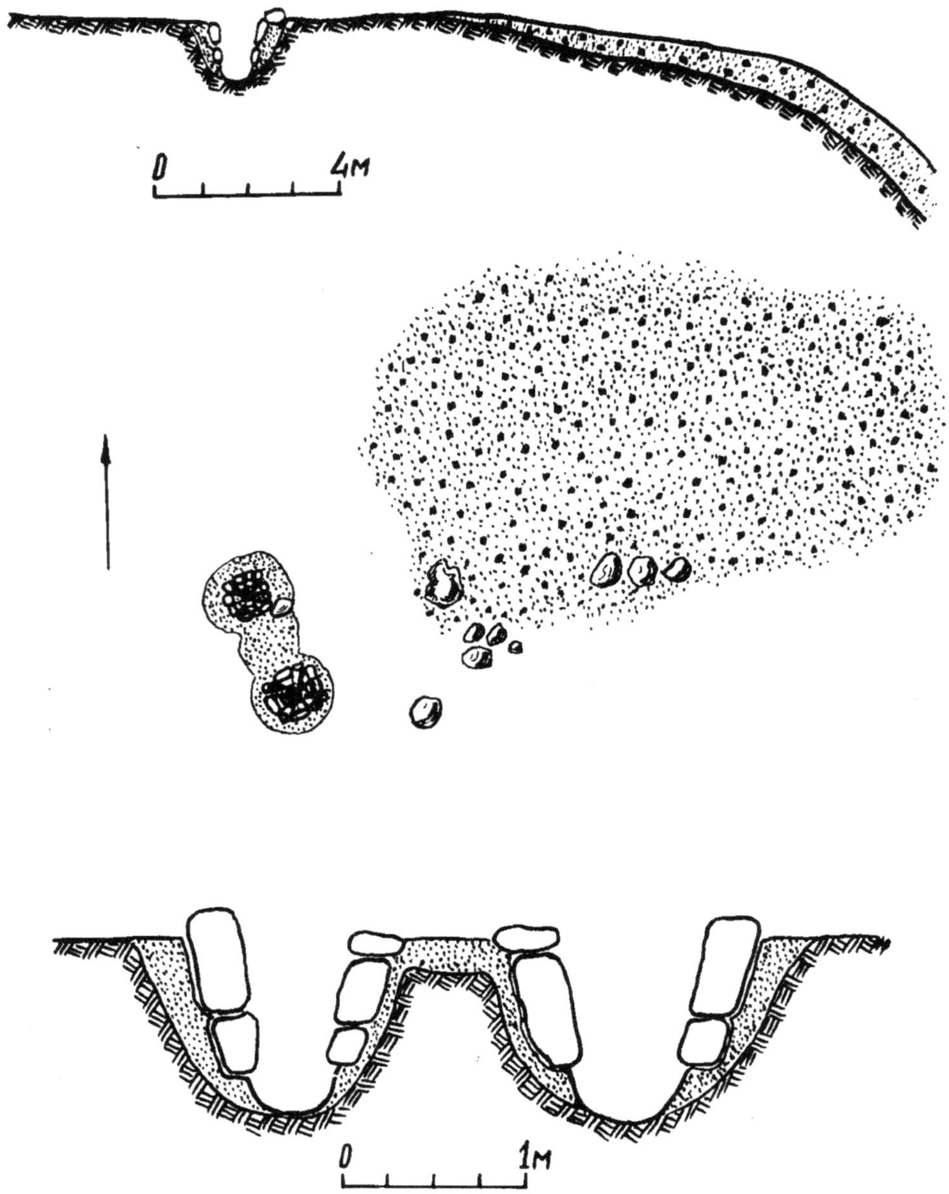

Fig. 4. Iron-Smelting Workshops "Charnali III." Ground Plan and Sections.

forge; in the second place, the archaeomagnetic date according to which the "Charnali III, 1" furnace in assessed point to the fact that "Charnali III,1" furnace is approximately 100 years older than "Charnali III, 2." Seeing that according to radiocarbon analysis "Charnali III, 1" is dated to the turn of the ninth and the eighth centuries B.C., the iron-smelting, workshop "Charnali III, 2" may be assigned to the eighth or seventh century B.C.

Thus, all the unearthed objects of the Chorokhi production centre including over 50 archaeological sites belong to the pre-antique period. The most ancient of them (Charnali I, 1, Charnali I, 2) functioned at the end of the second millennium B.C., and the latest in theeighth-seventh centuries B.C. Taking into consideration the significant role of the Chorokhi basin in the development of ancient metallurgy[22] and the role of South-Western Georgia in the formation of Colchian Late Bronze culture,[23] it may be stated that a certain part of the Chorokhi iron producing centre, a part that has not yet been excavated, belongs to an earlier period. This supposition is not unfounded: another centre of ancient iron manufacture situated further north contains much older iron-producing workshops (v. below, Chapter III). If we proceed from the universally accepted view that the earliest knowledge of iron smelting came to the Transcaucasus from the South, the logic of such a supposition will not appear to be unfounded. On the other

[22] A.A. Iessen. "On the Most Ancient Copper Metallurgy in the Caucasus." *Proceedings of GAIMK*, 120, Moscow, 1935, p. 33 ff.

[23] O.M. Djaparidze. "West-Georgian Culture in the Epoch of Late Bronze." *Matsné, Academy of Sciences of the GSSR, Historical Sources*, 1982, No.l, pp. 61-82; No.2, pp. 41-62 (in Georgian, summary in Russian).

hand, materials which we have at our disposal give us grounds to doubt that Transcaucasian masters, or Colchian masters in particular, were apprentices of Anatolian iron makers, or of those from any other region. We are bound to reiterate the opinion that we voiced well over twenty years ago that users of metal, especially of iron, are not necessarily producers of it.[24]

Consequently, an early mention of iron in written sources, or a discovery of it in an early material culture of a certain region does not necessarily mean that this metal was produced locally. On the other hand, the main producers of a metal are not always the principal consumers of it.

The masters working in the Chorokhi production centre used, as raw material, the magnetite they obtained on the sea shore. It has been proved that the beach sands in Western Georgia, from Gonio to Gagra and beyond contain a considerable quantity of magnetite (between 10 and 60 percent). The mineralogical composition of these sands shows them to be a mixture of magnetite grains, grains of titanium-magnetite, ilmenite, as well as fragments of olivine, hornblende and epidote.[25] A successful attempt was made to connect ancient settlements on the new Black Sea terrace (i.e. settlements in the dunes) with the acquisition of raw material for ancient iron metallurgy.[26] The use of magnetite for manufacturing iron is also corroborated by analysis of slags from ancient iron-smelting workshops. Settlements of this type have been identified along the entire coastline of Georgia, including the area south of the mouth of the Chorokhi River, on the territory of Apsarus (Gonio) fortress. As will be shown below, magnetitic sands were used in all centres situated near the sea, at distances from 5 to 20 km.

In the light of the materials of the Colchian iron-producing centre we may conclude that the notice ascribed to Aristotle concerning the Chalybean technique and raw material resources for the production of iron reflects a much earlier pan-Colchian reality that originated at some time in the second millennium B.C. In the words of Aristotle (or of his source?), "they speak of a quite peculiar origin of Chalybean and Amyss iron: it is obtained, they say, from sand borne by rivers; this sand, as some say, is simply washed and smelted, but according to others, the sediment remaining after washing is washed again several times and then smelted with an addition of the so-called refractory stone in which that country abounds. This sort of iron is much better than other sorts, and if it were smelted not in one furnace, it would be, it seems, in no way different from silver. As they say, it is only this iron that is resistant to corrosion, but it is obtained in small quantities."[27]

We should note here that the East Black Sea coast, all along the shoreline, is bordered by a swell of the Neopontic terrace which appeared in Late Holocene, in the period of the so-called Neopontic Transgression which began at about the middle of the 3rd millennium B.C. These processes of transgression were a tragedy for the population of the Colchian lowlands. They were forced to leave their homes and their coastal settlements owing to the sudden deterioration of economic conditions and a radical change of their living conditions.[28] The peak of development and the subsidence of the Neopontic Transgression cover the end of the third millennium and the first quarter of the second millennium B.C., i.e. the Middle Bronze Age. This was the time when the New Black Sea terrace took shape, and at the end of the second millennium B.C. a new kind of settlement arose, engaged, presumably, in obtaining the raw materials for making iron.[29] Such a settlement existed, as mentioned above, south of the River Chorokhi, between the estuary of this river and the village of Kvariati. The remains of this settlement are overlaid by strata of the Apsarus township site with its fortress belonging to Hellenistic and mediaeval periods. Similar settlements have been identified near all iron-manufacturing sites of the Colchian mining and metallurgical centre.

Judging by archaeological surveys, the Chorokhi centre of ancient iron manufacture consisted of a central area and its periphery. Most of the iron-smelting workshops were grouped in the central area, while single workshops were scattered over the peripheral area. Naturally, much had changed during the 2500 years of man's activity and certainly some of the iron-smelting workshops must have vanished irrevocably.

The same situation prevails in other centres of ancient Colchian metallurgy.

The region around the lower reaches of the Chorokhi, where the Chorokhi metallurgical centre is situated was populated in the Late Bronze and Early Iron Ages, by tribes belonging ethnically to the world of the Zeno-Kartic linguistic community. After the disintegration of this linguistic community, a considerable part of this region was occupied by a population that had infiltrated from territories inhabited by a Kart-speaking (East

[24] D.A. Khakhutaishvili. "At the Sources of Colchian Iron Metallurgy in the Caucasus", *Monuments of South-Western Georgia*, 1, Tbilisi, 1964, p. 50 (in Georgian).
[25] I.A. Gzelishvili. *Iron Smelting in Ancient Georgia*. Tbilisi, 1964, p. 16. D.A. Khakhutaishvili. "Raw Material Base and Technique of Iron Manufacture in Ancient Colchis". In: *History of Mining and Technology*. Tbilisi, 1977, pp. 86-88. A.T. Ramishvili. "On the Purpose of Sites with "Textile Ceramics" on the East Black Sea Coast." *Soviet Archaeology*, 1975, No.4, pp. 36-44.
[26] A.T. Ramishvili. *op.cit.*, pp. 36-44.

[27] Aristotle. *On Unbelievable Rumours*. 380B. T.S. Kaukhchishvili. *Greek Writers' Information about Georgia*, II, Tbilisi, 1969, pp. 67-68; idem: *Ancient Greek Sources on the History of Georgia*. Tbilisi, 1967, p. 68. (in Georgian).
[28] C.P. Djanelidze. *Palaeography of Georgia in the Holocene*. Tbilisi, 1980, pp. 21-64.
D.A. Khakhutaishvili. "Nature and Man on the Colchian Seaboard in the Holocene Period." *Caucasian-Near-Asian Collection*, Tbilisi, 1984, p. 146.
[29] Ch.P. Ramishvili. "On the Sites with "Textile Ceramics" on the East Black Sea Coast." *Soviet Archaeology*, 1975, No.4, pp. 36-44. See references in footnotes there.

Georgian) people. The territory we are investigating is a part of the littoral of historical Southern Colchis which had been inhabited by Kartvelian (Colchian, Zanian) tribes of the western group, including the Mossinicae and Chalybes. Then, the banks of the lower reaches of the Chorokhi are one of the principal maritime regions of South-Western Georgia connecting the northern regions of the Eastern Black Sea littoral with Asia Minor.

Presumably, a considerable part of the Chorokhi production centre is beyond the borders of modern Georgia. Farther to the South-West is the Chalybean centre famed in written sources and part of the ancient Colchian centre of iron.

The geographical location of the Chorokhi production centre and adjacent territories have given us grounds to suppose with a large degree of certainty that through it passed the main thoroughfare of trade, cultural and ethnic exchange between the central districts of the Eastern Black Sea littoral (Colchis) on the one hand, and Southern Georgia and Northern Anatolia on the other. The similarity of the technical equipment of iron production throughout the region can be explained by the cultural and ethnical identity of the population over the whole territory where this culture had spread. As O.M. Djaparidze has demonstrated, the littoral of South-Western Georgia was the main centre where the well-known Late Bronze Age Colchian culture, chronologically encompassing the Early Iron Age[30] took shape. It could have taken such a culturally important role only because it had deep roots in the cultural traditions of preceding epochs, which is convincingly corroborated by archaeological materials from the Georgian seaboard between the Chorokhi and the Supsa Rivers.[31]

[30] O.M. Djaparidze. "West-Georgian Culture of the Late Bronze Period." *Matsné, Series of History, Archaeology, Ethnography, and Art History*, 1982, No.l. pp. 61-82; No.2. pp. 42-62.

[31] D.A. Khakhutaishvili. "The Batumi Research Institute and Georgian Archaeology." In: *An Important Centre of Kartvelological Research*, Tbilisi, 1985, pp. 75-85 (in Georgian).

CHAPTER II

THE CHOLOKI-OCHKHAMURI MANUFACTURING AREA

The Choloki-Ochkhamuri manufacturing area of ancient iron metallurgy is situated in a zone along the middle course of the rivers Choloki and Ochkhamuri, on the territory of the Kobuleti and Ozurgeti districts, Georgian SSR (Fig. l).

To the north of Kobuleti seaside resort, on the territory of pre-Antique settlements overlaid by strata of a nameless antique town, the rivers Choloki and Ochkhamuri flow together. The basins of these two rivers are limited, on the West by the Black Sea; on the East by spurs of the Adjarian-Gurian mountain range descending to the sea like an amphitheatre; on the South by the gorge of the Kintrishi river, and on the North by the Natanebi river gorge.

This manufacturing area is situated at a distance of 40-50 km. from the Chorokhi area. The Choloki and Ochkhamuri rivers flow across the southernmost part of the Colchian alluvial valley cut off from the principal massif of the Colchian accumulative valley by a chain of hills coming down to the Black Sea.

The central part of this manufacturing area lies on both banks of the Choloki, near Djikhandjuri and Tsetskhluri villages (Kobuleti district), on the land of the Anaseuli and Narudja state farms and of the village of Makvaneti. The peripheral iron-smelting workshops are scattered on the territories of the villages: Leğva, Kakuti, Ekadiye, Natanebi, Kviriké, etc. (Fig. 4A).

On this territory, with its thick network of gullies draining into the Choloki river or the Ochkhamuri, there are thick strata of red soils (laterites), resulting from the erosion of various tufogenic formations of the Middle Eocene, overlaid by a layer of deluvial yellowish clays. In this region red soils have a structure typical of clayey formations to a depth of 3 metres. Lower down, from 3 to 20 metres and deeper, red soils in places have the structure of the matrix.[1] On this territory there are abundant deposits of fire-clay known in specialist literature at the "Tsetskhlauri fire-clays;"[2] at a distance of 10-15 kilometres from the centre of this manufacturing area are the richest deposits of magnetitic sands[3] on the Eastern Black Sea littoral. Magnetitic sands carried by rivers from mountain slopes (recall Aristotle's report) are the result of natural erosion of iron-bearing rock, mainly magnetite. These sands are carried into the sea and then washed by the breakers along the beaches, forming layers more or less rich in magnetite. As some researchers conjecture, the first terrestrial iron was smelted from magnetitic sand. The work done on the Eastern Black Sea littoral corroborates this. We can only regret that at present we are as yet unable to give an unambiguous answer as to how this actually took place. But we think it would be quite natural to consider that ancient man first gained an acquaintance with "native" iron, and only later began to experiment in attempts to evolve the technique of obtaining "terrestrial" iron.

The zone occupied by the iron-smelting workshops of the production area in question was covered with thick forests composed of every kind of subtropical plant, characteristic of a humid climate. The local forest massifs provided for the production of any kind of charcoal required for bloomery furnaces. Thus the area around the middle reaches of the Choloki and Ochkhamuri rivers offered every condition for the initiation of iron manufacture: magnetites, fire-clays and good fuel and, with the existing culture and traditions of metal manufacture, led to the creation of a large and long-lasting metallurgical centre which operated from the latter half of the second millennium B.C. to the sixth century B.C. or, possibly, to a still later date.

The first traces proving the existence of ancient iron manufacture on the territory around the middle reaches of the Choloki and Ochkhamuri rivers were identified in 1953, but the excavation of the site and the archaeological investigation of the territory began in 1960.[4]

Investigation of the site and the area over which sites of ancient iron production were spread showed that the Choloki-Ochkhamuri centre contained about 100 sites that were mainly grouped on the land of the Tsetskhlauri, Djikhandjuri, Anaseuli and Narudja state farms, while separate workshops were scattered over a much greater territory (villages; Leğva, Kakuti, Kviriké, Ekadiye, etc.). Iron-smelting workshops of the Anaseuli and Narudja group were on the right bank of the Choloki, between which river and the Ochkhamuri were the Djikhandjuri and Tsetskhlauri groups.[5]

[1] I.A. Gzelishvili. *Iron Manufacturing in Ancient Georgia*. Tbilisi, 1964, p. 38.

[2] Idem, *Mineral Resources of the Georgian SSR*. Tbilisi, 1935, pp. 160-167. Gzelishvili. *op.cit.*, p. 39.

[3] Idem, pp. 244-247. *Natural Resources of the Georgian SSR*, I, Moscow, 1959, p. 83. Gzelishvili. *op.cit.*, p. 39.

[4] During my post-graduate studies at Tbilisi State University, I worked in 1953 on the littoral of South-Western Georgia collecting factual data for my master's thesis, "Kartvelian Tribes of the Black Sea Littoral According to Greek and Latin writers." In 1953, R. Khvichia, a villager from Mukhaĕstaté, a village on the territory of Tsetskhauri state farm in the Ontapura area, turned up several dumps of iron slag, about which we informed the archaeological expedition of the Institute of History, of the Academy of Sciences of the GSSR, then working on the site of the settlement in Pichvnari.

[5] Cf. Gzelishvili. *Iron Smelting in Ancient Georgia*, p. 38 ff.

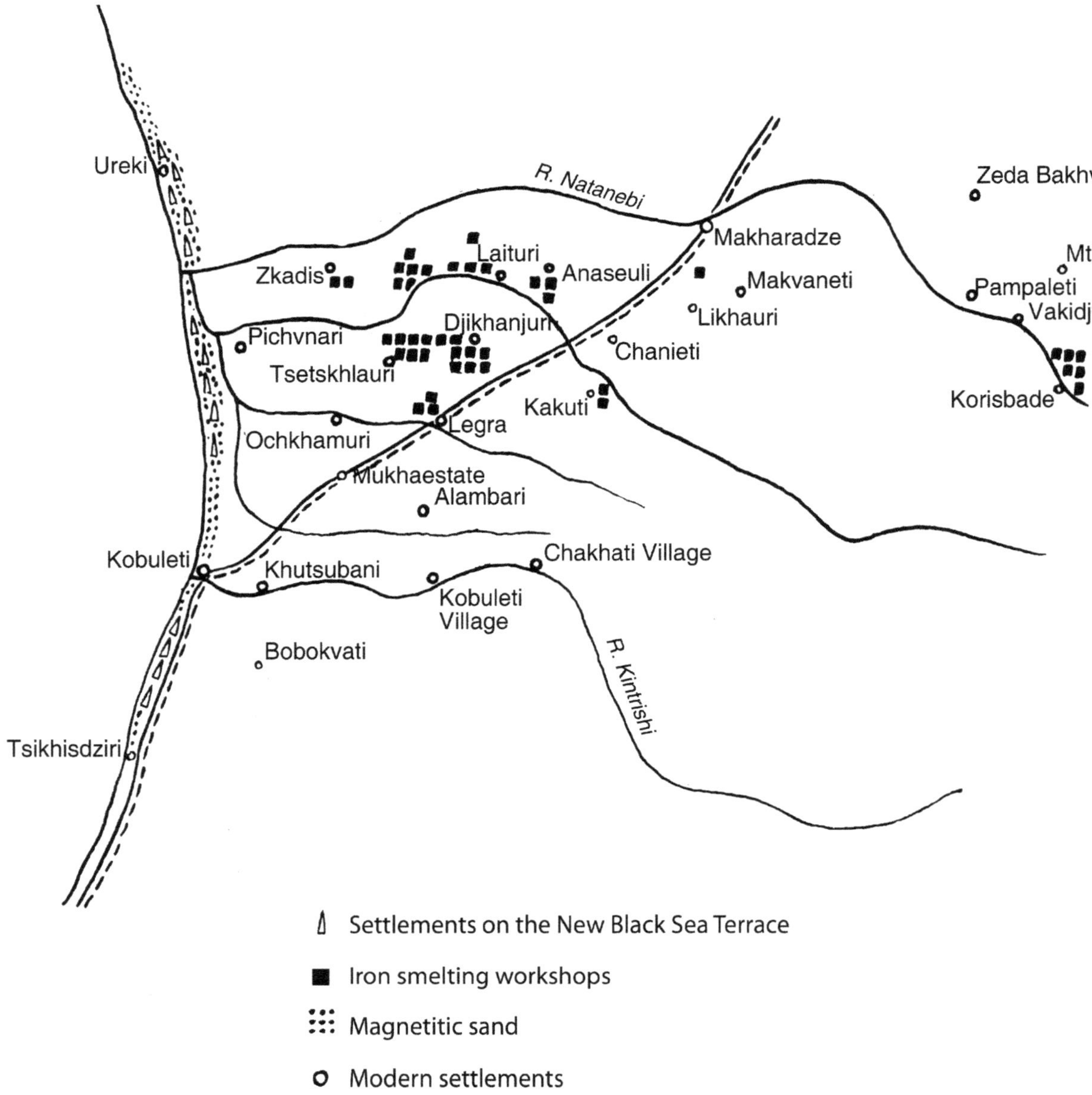

Fig. 4A. Location of separate groups of iron-smelting workshops of the Choloki-Ochkhamuri production centre.

At present, eight archaeological sites have been unearthed, four of them of the Djikhandjuri group (excavated by I.A. Gzelishvili), three of the Tsetskhlauri group (excavated by D.A. Khakhutaishvili) and one site in the village of Leğva (excavated by I.A. Gzelishvili).[6]

1. IRON-SMELTING WORKSHOPS "DJIKHANDJURI I"

This archaeological site[7] containing three iron slag dumps was excavated in 1960 by I.A. Gzelishvili. It is situated on the territory of the Djikhandjuri state tung-tree plantation (used in the production of soap, Kobuleti district, Georgian SSR), on a smallish promontory-like area at the edge of a deep gully, near the "Chapayev shed." Under the trees there is a growth of low fern.

The remnants of the iron-smelting furnaces were covered with a layer of cultivated soil 30 to 40 cm thick. Beneath this layer three iron-smelting furnaces were revealed, as well as the remnants of three stone tables (anvils) for forging the spongy mass extracted from the furnace, and three waste dumps and materials which will be dealt with below.

[6] Gzelishvili. *Iron Smelting in Ancient Georgia.* pp. 38-47. D.A. Khakhutaishvili. *Newly Discovered Sites of Ancient Colchian Iron Metallurgy.* 1977, pp. 29-33.

[7] The Djikhandjuri iron-smelting workshops were excavated In 1960-1961 by a joint archaeological expedition of the I.A. Djavakhishvili Institute of History, Academy of Sciences of the Georgian SSR, and the Batumi Research Institute of the Georgian Academy of Arts. Head of the expedition I.A. Gzelishvili; scientific consultant Academician of the Georgian SSR Academy of Sciences Prof.N.A. Berdzenishvili; deputy head of the expedition D.A. Khakhutaishvili.

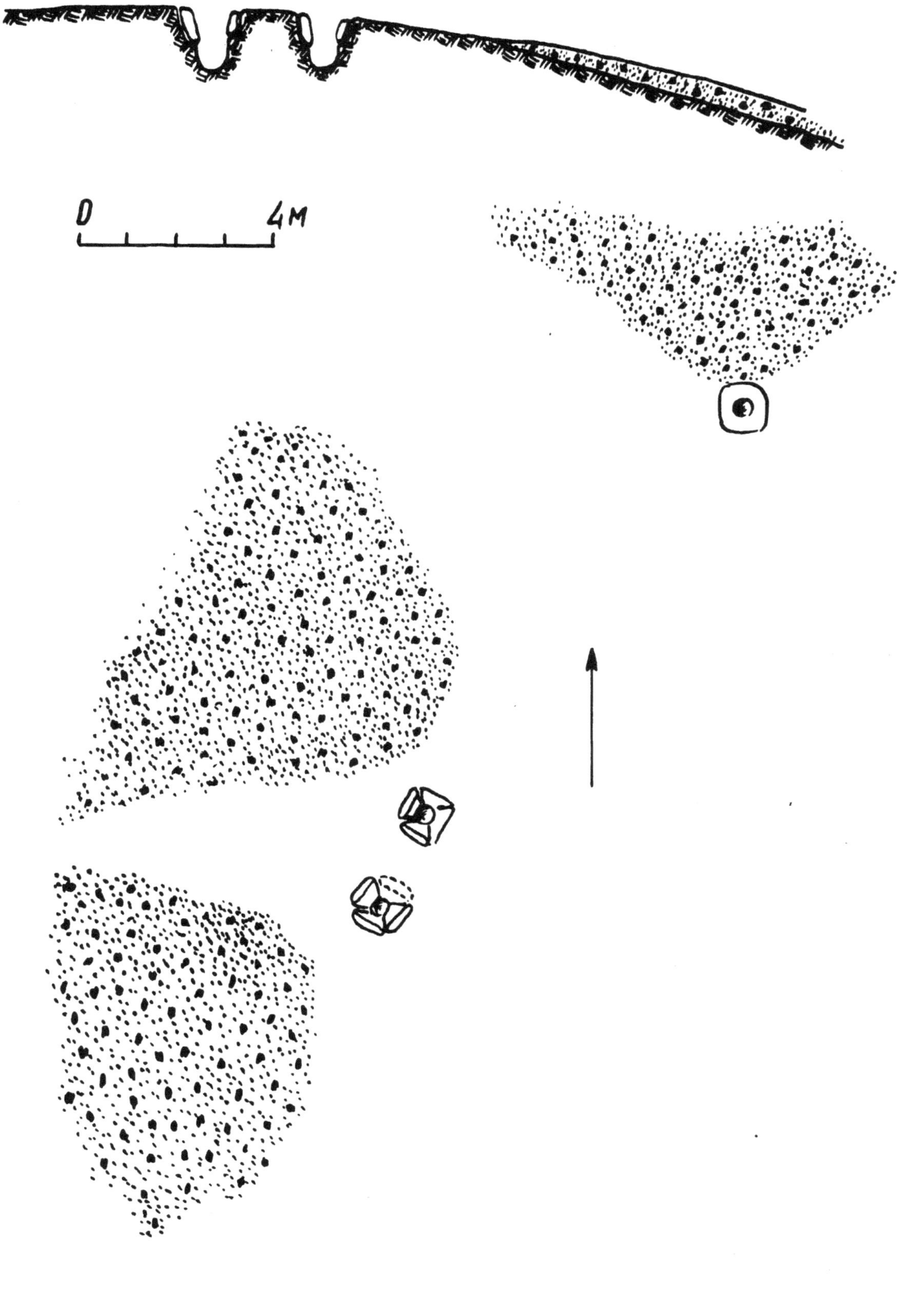

Fig. 5. Iron-smelting workshops "Djikhandjuri I." General ground plan and section.

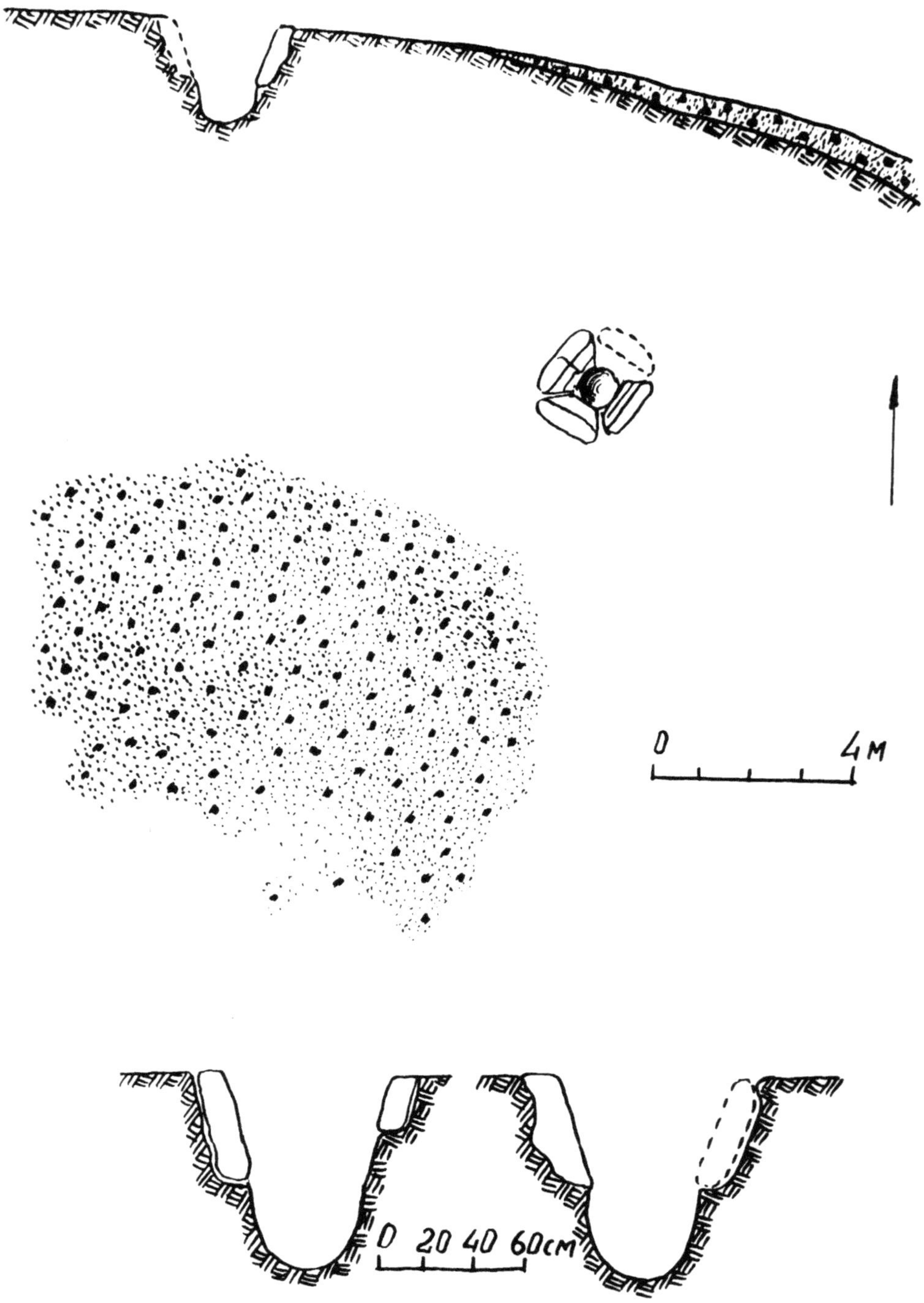

Fig. 6. Iron-smelting workshop "Djikhandjuri I, 1." General ground plan and sections.

a. Iron-Smelting Furnace "Djikhandjuri I, 1"

The iron-smelting workshops "Djikhandjuri I, 1" consisted of the iron-smelting furnace, remnants of the stone table for working the bloom, the waste dump and other materials found when investigating separate parts of the archaeological site (Fig. 5).

The iron-smelting furnace "Djikhandjuri I, 1"[8] (Fig. 6) has survived in the form of a pit shaped like an upturned truncated pyramid. The upper two-thirds of the pit are lined with stone, and the lower one-third is coated with a thick layer of fire-clay; evidently, the whole inner surface of the furnace was plastered in this way. The stones of the facing and the plastering of the lower part of the pit are well preserved.

In the walls of the surviving part of the furnace there are no orifices for the nozzles of the bellows, although several fragments of such nozzles of clay were found in the waste dump.

[8] In 1960 this furnace was conserved *in situ*. See: .I.A. Gzelishvili. *Iron Smelting in Ancient Georgia*, pp. 39-41, Figs. 9A, 9B, B, pl. IX A.

The surviving part of the furnace is 90 cm high, the width at the upper rim is 85 cm, in the middle 55 cm, at the bottom 10-15 cm. The ground around the furnace shows signs of burning at a high temperature.

Inside, the furnace was filled with clay mixed with iron slag pieces, melted fragments of the clay lining and otheradmixtures. At the bottom of the furnace was a layer of roughly ground charcoal mixed with clay.

Under the impact of intense heat, the inner surface of the furnace was coloured mauve in places which testifies to a considerable power potential at the disposal of the master smelters.

The table (anvil) for forging the bloom has survived only in part as an area burnt red and containing a few flat stones, red at fractures, most probably ascribable to the high temperatures.

The waste dump, 5 metres to the west of the furnace, covered about 30 m^2 and was 50-60 cm thick in the centre.

The dump contained a great many lumps of iron slag (totalling almost 5 cubic metres) and entire round "cakes", or truncated conical features (upper diameter 30-50 cm, bottom diameter 20-25 cm, 8-15 cm high). On some of these "cakes" pieces of fire-clay plastering had survived, fused in the process of smelting. It is noteworthy that the dull-grey "thick" lumps of slag have patina pointing, as I.A. Gzelishvili rightly remarks, to their antiquity. Besides slag, the dump contains a great many fragments of fire-clay plastering. Judging by their quantoty, the furnace must have been re-plastered before every new smelting.

The volume of slag was about 4 m^3 and that of plastering 1 m^3. Near the top of the dump a flat plot was revealed where fire-clay was stored: on a 4 m^2 area, a mass of yellowish-grey malleable clay over 20 cm thick was prepared for further use.

I.A. Gzelishvili did not dwell upon the problem of dating this workshop as he lacked the relevant archaeological material; but, taking into consideration the general situation and the finds from other iron-smelting workshops excavated by him in 1969 at Djikhandjuri, he opined that "the seventh century B.C. should be considered the latest date of iron smelting at Djikhandjuri."[9]

In the 1970s, Z.A. Chelidze carried out an archaeomagnetic investigation of the excavated remains of ancient iron manufacture and came to the preliminary conclusion that the iron-smelting furnace "Djikhandjuri I, 1" belongs to the ninth century B.C. (declination = -8.0°, inclination = 74.0°, tension = 0.353 oersted).[10]

It should be mentioned here that the "Tsetskhlauri" groups of iron-smelting furnaces dated to the ninth century B.C. served as a basis for the geophysical dating of the above.[11]

b. Iron-Smelting Workshop "Djikhandjuri I, 2"

The "Djikhandjuri I, 2" iron-smelting workshop was excavated in 1960. The following parts have survived: the iron-smelting furnace, remnants of the table (anvil) for working the bloom and a flat space for the waste dump (Fig. 7), as well as other materials revealed when clearing the archaeological stratum.

The iron-smelting furnace "Djikhandjuri I, 2"[12] is a pit shaped like an upturned truncated pyramid. The upper two-thirds of the pit were faced with stone, The lower one-third was plastered with a thick layer of fire-clay. It may be surmised that all the inner surface of the furnace was plastered in the same way. The facing stones and the plastering of the lower part of the furnace are in a good state of preservation.

The walls of the surviving part of the furnace have no nozzle notch for the bellows, although the preparation of the dump yielded fragments of cone-shaped nozzles of bellows.

Inside, the furnace was filled in with clay mixed with lumps of iron slag here and there fused with the furnace wall, melted fragments of the plastering, and other alien admixtures. At the bottom was a layer of rough-ground charcoal mixed with clay.

The height of the surviving part of the furnace was 90 cm, the width at the top line 65 cm; in the middle 50 cm, at the bottom 10-15 cm. The ground round the furnace was burnt red to a depth of 15-20 cm.

The inner surface of the furnace was purplish owing to the high temperature it had been subjected to.

Clearing the furnace yielded: the loop-shaped handle of an earthenware vessel, as well as sherds of the walls of large red earthenware pots with vertical stripes (Fig. 16, 1-3).

Of the table (anvil) only fragments remained. It was probably destroyed when the plot was cultivated before

[9] I.A. Gzelishvili. *op.cit.*, p. 51

[10] Z.A. Chelidze. Results of Archaeomagnetic Investigation of some Archaeological Sites. *Problems of Ancient History* (Caucasian-Near-Eastern Collection, V, Tbilisi, 1977, pp. 142-145.

[11] D.A. Khakhutaishvili. "Excavation of an Ancient Colchian Centre of Iron Manufacture in the Valley of Choloki in 1971." *Matsné*. (in Georgian); Tbilisi, 1974. p. 202; Idem. Materials for the History of Ancient Colchian Iron Metallurgy. (Ochkhamuri Valley). In: *Sites of South-Western Georgia*, VII, Tbilisi, 1978, pp. 32-38 (in Georgian, summary in Russian).

[12] After being cleared and studied, the furnace was conserved in 1960.

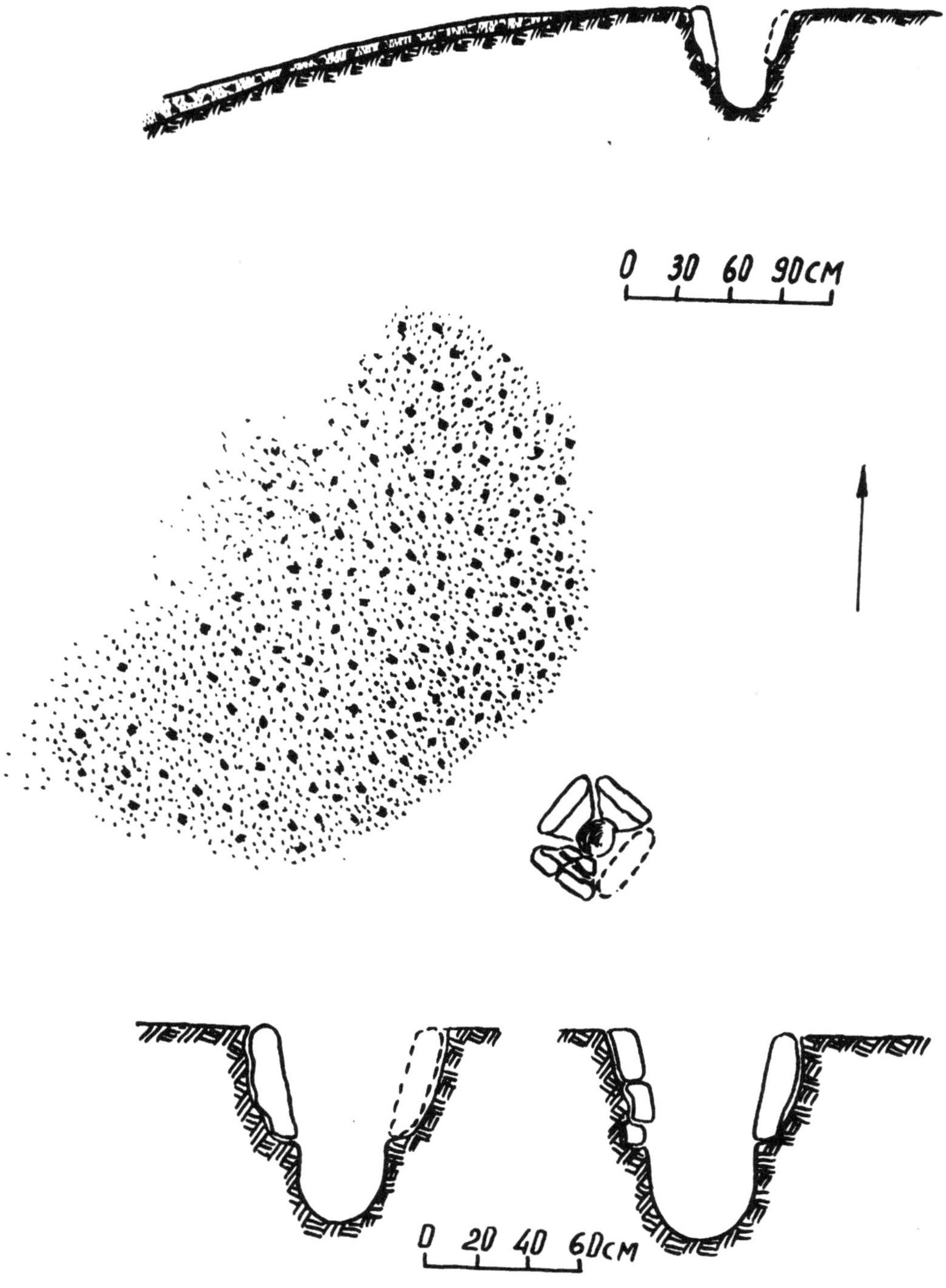

Fig. 7. Iron-smelting workshop "Djikhandjuri I, 2." General ground plan and sections.

planting tung trees. The area burnt red by the heat was cleared and separate stones were found in the same condition.

The waste dump three to four metres north-west of the furnace covered about 20 m^2 of space. The thickness of the dump is 30 to 40 cm.

The waste contained: much iron slag (about 4 cubic metres), the lumps being covered with so-called "thick" patina, dull-grey in colour, fragments of the clay plastering of the furnaces, a whole clay "cake" and pieces of clay nozzles for the bellows, ash, pieces of charcoal, stones of different shapes including some burnt red, etc. Closer to the furnaces, under a layer of ashes, a dump was cleared containing yellowish-grey plastic fire-clay.

The iron-smelting workshop is dated to the eighth-seventh centuries B.C. This dating is supported by sherds of pottery unearthed in the clearing of the iron-smelting

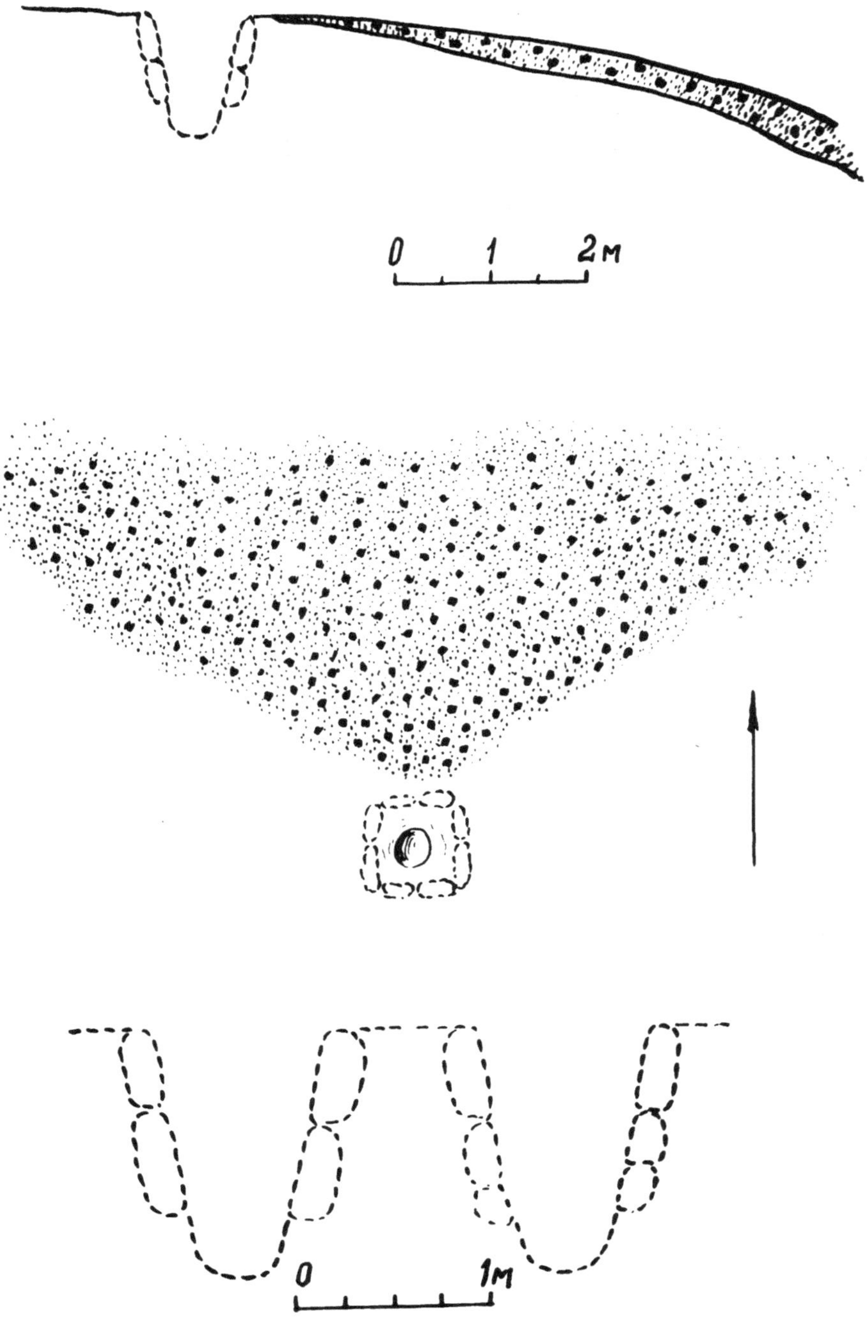

Fig. 8. Iron-smelting workshop “Djikhandjuri I, 3.” General ground plan and sections.

furnace.[13] They have analogues in materials from the so-called dune settlements on the new Black Sea terraceOn the other hand, according to the archaeomagnetic investigation by Z.A. Chelidze, the last smelting in the “Djikhandjuri I, 2” furnace was effected at about the middle of the eighth century B.C. (declination = +4.0°, inclination = 63.5°; tension = 0.474 oersted).[14]

[13] Gzelishvili. *op.cit.* pp. 41, 49-50, pl. IX, A, 5-7.

[14] Z.A Chelidze. *op.cit.*, pp. 143-145; idem. “Results of Archaeomagnetic Investigation of Some Iron-Smelting Workshops Identified on the Territory of Ancient Colchis” (in Georgian, summary in Russian). In: *Sites of South-Western Georgia*, VII, Tbilisi, 1978, pp. 35-41.

c. Iron-Smelting Workshop “Djikhandjuri I, 3”

The iron-smelting workshop “Djikhandjuri I, 3” was excavated in 1960.

The iron-smelting furnace, remnants of the table (anvil) for forging the bloom and the waste dump have been identified (Fig. 8).

The iron-smelting furnace “Djikhandjuri I, 3”[15] is a pit, formed like an upturned truncated pyramid. The facing of the furnace has survived only in the lower part, namely,

[15] The furnace has been conserved.

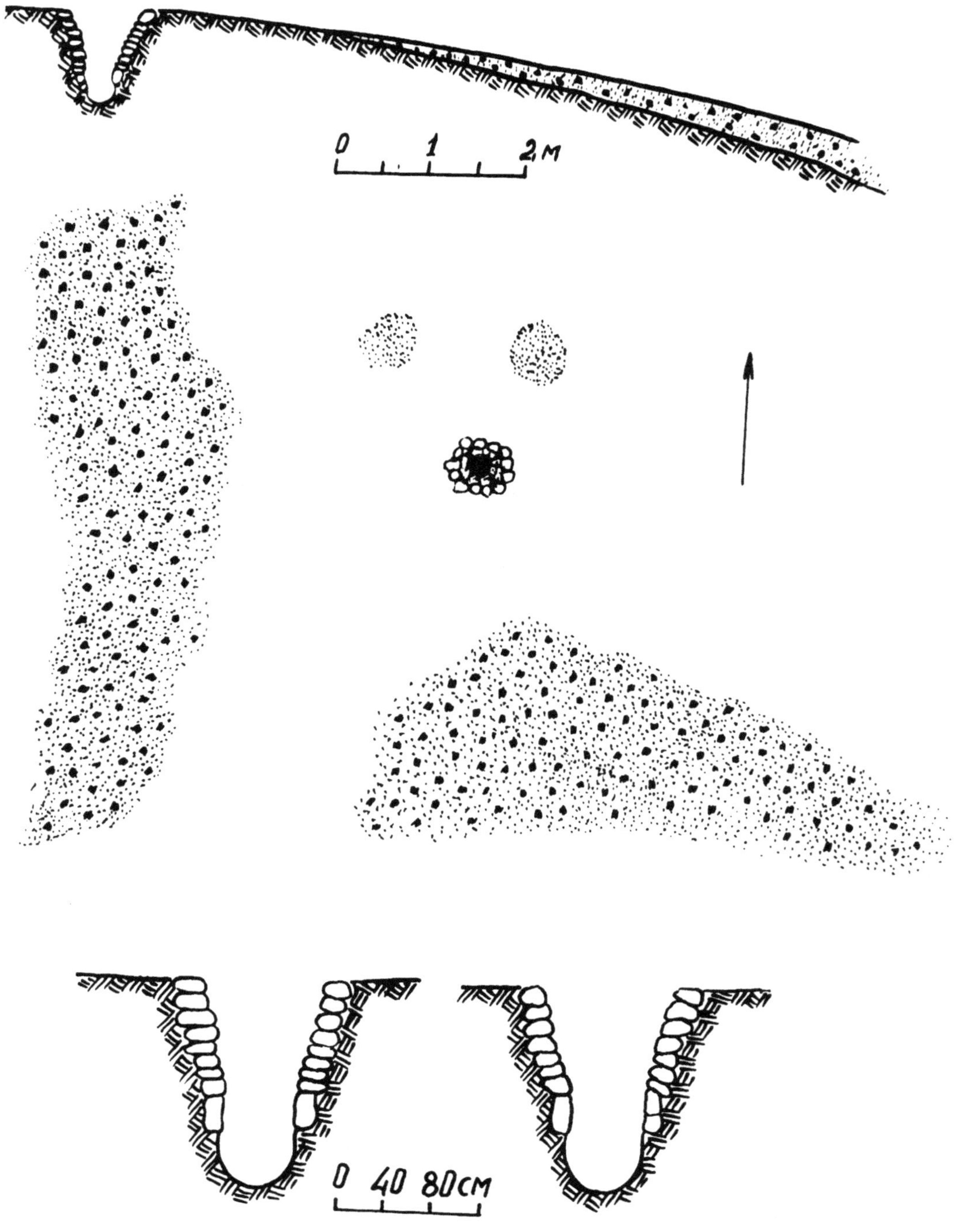

Fig. 9. Iron-smelting workshop "Djikhandjuri II." General ground plan and sections.

the thick coating of fire-clay has been preserved. The facing stones of the upper part of the pit are absent (Fig. 8). The inside of the pit was filled in with clay and various materials: the facing stones that had fallen in and were reddish at fractures, lumps of slag, bits of charcoal, etc. The whole perimeter of the pit was reddish - the result of high temperatures. At the bottom was a layer of roughly ground charcoal mixed with clay.

Of the table (anvil) for working the bloom, there remained only a strip of ground burnt red and a few scattered stones, reddish at fractures (resulting from the intense heat of the spongy mass being forged).

The waste dumps lying to the north of the furnace at a distance of 4-5 metres covered a space of about 40 m^2. The thickness of the layer in the centre of the dump was about 40 cm.The waste contained a great deal of iron slag (up to 5 m^3), fragments of fire-clay plastering of the iron-smelting furnaces, fragments of clay nozzles of bellows, "singed" stones, ash, etc. At the edge of the dump, closer to the furnace, was an unused quantity of fire-clay. The lumps of iron slag have the characteristic dull-grey "thick" patina.

Concerning the age of this workshop, we might state the following:

1. According to the structure of the smelting furnace and aspect of the slag, the workshop belongs to an early period;
2. According to the archaeomagnetic data of this workshop obtained by Chelidze, the last time iron was smelted in it was about the end of the ninth century B.C. (declination = +28.0°, inclination = 71.5°, tension = 0.590 oersted).[16]

Besides the materials mentioned above, the excavation of the "Djikhandjuri" site yielded number of objects during clearing work on the upper area where the workshops were located: a stone mortar for pounding fluxes, a fragment of a clay nozzle with a notch and a number of other fragments of clay objects of unidentified purpose, most probably the nozzle for bellows.[17]

An analysis of slags from "Djikhandjuri I" has shown that during smelting ore the temperature rose to 1200-1250° C.[18] The slags consist mainly of fayalite and magnetite, the interstices being filled with glass. This is due to the fact that the slags from Djikhandjuri are the product of smelting magnetic sand.[19]

Thus, proceeding from the materials obtained at "Djikhandjuri I", we can assert the identity of industrial skills and technical equipment of iron manufacture in the Choloki and Choloki-Ochkhamuri iron manufacturing centres.

"Djikhandjuri I" with its three iron-smelting workshops is so far the only one of its kind among the sites we have unearthed.

2. IRON-SMELTING WORKSHOP "DJIKHANDJURI II"

The iron-smelting workshop "Djikhandjuri II" (Fig. 9)[20] was excavated by I.A. Gzelishvili in 1960.[21] It is situated 500 metres to the north-west of the iron-smelting workshops "Djikhandjuri I", on the top of a hillock. The remnants of the workshop were identified at a depth of 30-40 cm beneath a layer of yellowish clay and topsoil. The waste dumps were on the north and south slopes of the hill. The state farm had planted tung-trees on the territory of the workshop. *Rhododendron ponticus* and other species had grown in hollows. The whole area beneath the tung-trees was overgrown with low fern with a deep-penetrating root system.

[16] Chelidze. *op. cit.*, p. 145.
[17] Gzelishvili, *op. cit.*, pp. 49-51, pl. IX, A.
[18] Ibid., p. 83.
[19] Ibid., p. 82.
[20] Gzelishvili. *op. cit.*, pp. 41-43. Fig. 10, pl.1, 4, IX, B, 4-9.
[21] Ibid.

The iron-smelting furnace "Djikhandjuri II" (Fig. 9)[22] is a pit shaped like an upturned truncated pyramid ending in a spherical hollow. The facing of the furnace is in good condition. The lower part of the pit is plastered with a thick layer of fire-clay to a height of 35 cm, while the top part (95 cm) is lined with stone. An interesting point is that the lower part of the stone-lined portion consists of four slabs of rough flat stone 30 cm high, and the top part (65 cm) of rough quarry stone.

The furnace is 130 cm high, 80 cm wide at the top, 45 cm wide in the middle and 10-15 cm at the bottom. The diameter of the spherical hollow is 45 cm at the top.

The whole interior of the furnace was filled in with yellowish clay mixed with remnants of charcoal and ashes, stones and lumps of slag. The rounded bottom of the furnace was, as in all the cases described above, filled with crushed charcoal with an admixture of clay. In the upper part of the furnace bits of iron slag were found, fused with the wall. Among the material found when the furnace was being cleared were fragments of bellow nozzles.

Of the table (anvil) for working the bloom only some parts have remained; to the north-west of the iron-smelting furnace, at a distance of two metres, two patches of yellow diluvial clay burnt red were discovered. Around them were found prepared flat stones, possibly serving as supports for a wooden shed.[23]

Waste dumps were on the northern and southern slopes of the hill. They contain a great many lumps of iron slag and pieces of clay plastering of the furnace, as well as fragments of nozzles for bellows.[24] All this material was mixed with yellowish clay, ash and small pieces of charcoal.

On the north slope, at the top edge of the dump under a layer of slag an area (about 4 m^2) was cleared, with a stock of fire-clay 20 cm thick. Clearing yielded fragments of ceramic vessels on the upper part of the furnace: two loop-shaped handles with spool-shaped attachments, and sherds of a large burnished vessel of red clay[25] (Figs. 19, 12, 13).

The "Djikhandjuri II" workshop is dated to the eighth-seventh centuries B.C. This date is supported by the pottery which is analogous to the material from the so-called dune settlement in Pichvnari dated to the eighth-seventh centuries B.C.[26] Seeing that the pot handles from "Djikhandjuri II" belong to later variants of such objects,

[22] The iron-smelting furnace was cut out as a whole and is exhibited in the State Museum of the Adjarian ASSR in Batumi.
[23] Gzelishvili. *op. cit.*, p. 41.
[24] A.T. Ramishvili. *Excavation of Seaside Settlements in Pichvnari*, pp. 136-153, figs, 10, 7-10. Idem: *On Dating the Seaside Settlements in Pichvnari*, pp. 104-110, pl. V.
[25] Gzelishvili. *op. cit.*, pp. 43-44, fig. 11, pl. IX, B, 1-3.
[26] Gzelishvili. *op. cit.*, pl. IX, B, 6-7.

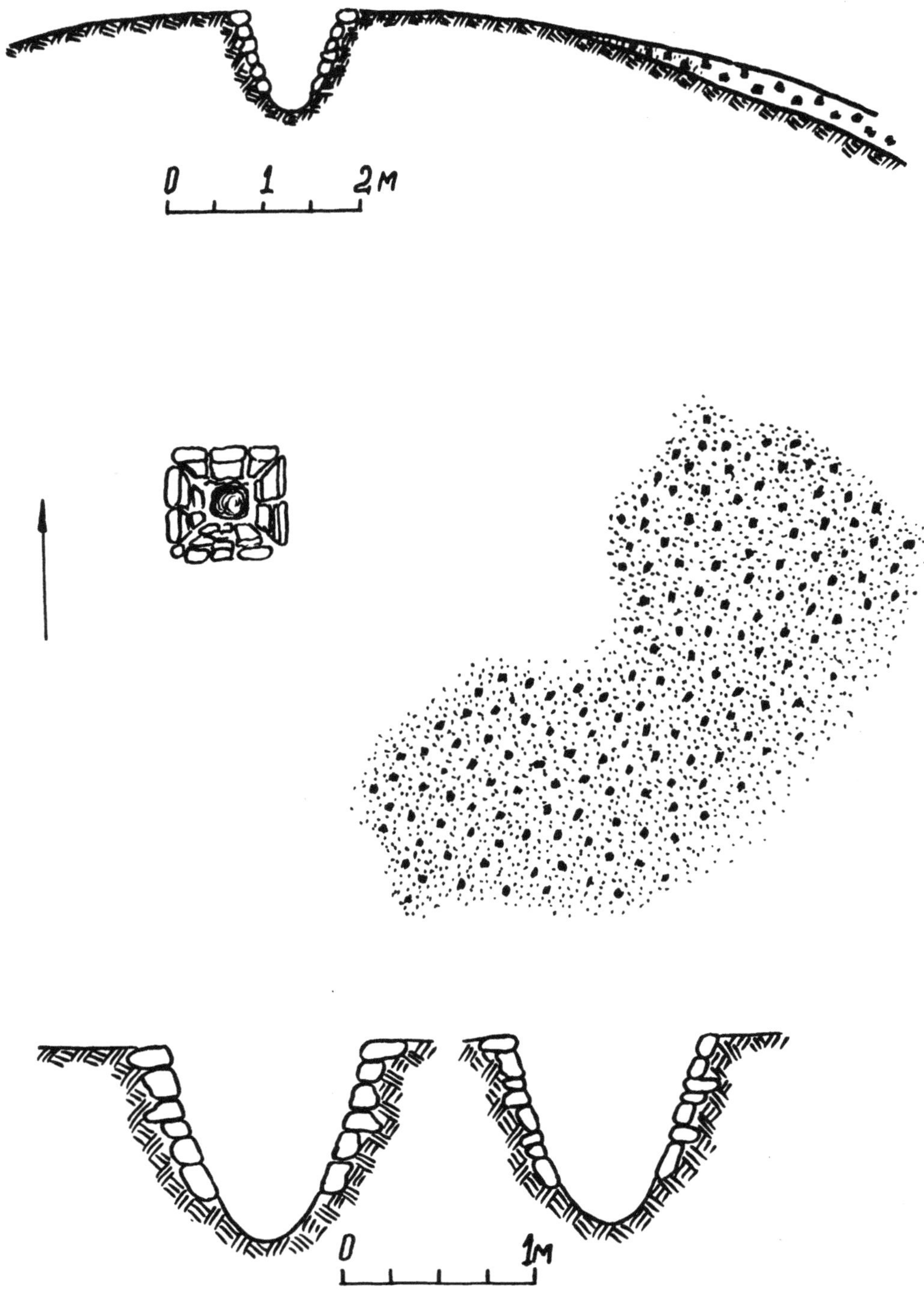

Fig. 10. Iron-smelting workshops "Djikhandjuri III." General ground plan and sections.

we are inclined to date the "Djikhandjuri II" workshop to the seventh century B.C.

3. IRON-SMELTING WORKSHOP "DJIKHANDJURI III"

The iron-smelting workshop "Djikhandjuri III" lies to the west of the "Djikhandjuri I" workshops, at a distance of 500 metres from them, on a promontory-like hill stretching southwards and bordered by gullies on the west and the east. This workshop was excavated by I.A. Gzelishvili[27] in 1960. Remnants of the workshop were found at the top of the hill, under the topsoil and a layer of yellowish clay 30-50 cm thick.

Like all the workshops described above, this one consisted of the iron-smelting furnace, the table (anvil) for forging the bloom and the waste dump.The iron-smelting furnace "Djikhandjuri III" (Fig. 10) was unearthed at the top of the above-mentioned hill, at the western edge of the area. The furnace was an upturned

[27] Gzelishvili. *Iron Smelting*. pp. 43-44; Fig. 11, pl. IX, B, 1-3.

truncated pyramid in form, the top part of which was faced with stone; the lower part, a hemispherical hollow, was plastered with a thick layer of fire-clay. When in working condition the entire inner surface of the furnace was smeared with fire-clay.

The interior of the furnace was filled in with clay and humus mixed with lumps of slag, fragments of the clay plastering, of the bellow nozzles, ash and charcoal. At the bottom of the furnace was a layer of roughly crushed charcoal mixed with clay. Inside the furnace there was found the smooth loop-shaped handle of a red earthenware pot (Figs. 19, 10, 11).

The furnace was 90 cm high; width at surface 90 cm, in the middle 40 cm, at the bottom 5-10 cm. Owing to intense heat, the earth around the furnace was reddish, merging with yellow clay at the edges.

The table (anvil) for forging the bloom was 2 metres from the furnace. There remained a patch burned red, and another near it, where, probably, the hot mass from the furnace was dumped. The stones which composed the table were scattered over the whole area occupied by the workshop. It may be conjectured that the table was dismantled when the ground was being prepared for the tung-tree plantation.

The waste dump was on the east slope of the hill. In it were numerous lumps of slag (totalling about 5 cubic metres), fragments of the clay plastering of the furnace, as well as nozzles for bellows. The dump covered 40 m^2; its maximum depth was 30-40 cm. A stock of fire-clay, yellowish-grey in colour, was cleared at the upper edge of the waste dump.

The date of this workshop may be determined as: eighth-seventh centuries, seeing that the earthenware vessel from the furnace has an exact analogue in the dune settlements at Pichvnari, dated, as mentioned above, to the eighth-seventh centuries B.C.

4. THE IRON-SMELTING WORKSHOP "DJIKHANDJURI IV"

The iron-smelting workshop "Djikhandjuri IV" (Fig. 11) is situated to the west of the workers' settlement of the Djikhandjuri tung-tree state farm, at a distance of about 1 km., in the farm's section No.2, near shed No.1. It was excavated by I.A. Gzelishvili in 1961.[28] The platform for the workshop was made on the steep bank of a nameless gully. The whole slope, before the excavation began, was planted with tung-trees, with a thick growth of plants beneath them.

The platform, about 140 m^2, on which the iron-smelting shop was arranged, was covered with a layer of topsoil and diluvial yellow clay under which, at a depth of 50-70 cm, began the top horizon of the remnants of the workshop.

The "Djikhandjuri IV" iron-smelting workshop consisted of the iron-smelting furnace, the table (anvil) for forging the bloom and the waste dump.

The furnace "Djikhandjuri IV" (Fig. 11)[29] was situated in the north part of a levelled plot. The furnace is a pit shaped like an upturned truncated pyramid, penetrating into a stratum of Tertiary tufo-breccia. The top part of the furnace was faced with stone (roughly hewn slabs); the lower part ending in a rounded hollow was plastered with fire-clay (of which fragments remain). The interior of the furnace was filled in with lumps of slag, clay plastering and nozzles of bellows. All this material was mixed with clay, ash and pieces of charcoal. At the bottom was a layer of roughly crushed charcoal smeared with clay. Around the furnace, the ground was burned red.

The dimensions of the surviving part of the furnace are: height - 130 cm; width: at the rim - 110 cm, in the middle - 45 cm, at bottom - 10-15 cm The depth of the rounded hollow is 25 cm.

The table (anvil) for working the bloom had partly survived as a stone pavement less than 2 m^2. The stones affected by the great heat were red at fractures.

The waste dump was on a slope opposite the furnace to the north of the latter and covered an area of about 60 m2 The maximum depth of the cultural stratum was 40 cm. The dump contained lumps of iron slag (up to 10 cubic metres), fragments of clay plastering and bellows nozzles, stones of various shapes, ash and occasional pieces of charcoal. All this material was mixed with clay and topsoil.

After clearing the iron-smelting furnace, a large pit was dug on its east side; the bottom of this pit was dug 50 cm lower than the rounded hollow of the furnace. Then the east wall of the furnace and of the pit housing it was dismantled. It was found that the stone masonry of the interior of the furnace rested on a narrowish ledge left around the pit. It also turned out that the stone walls of the furnace were permanent. I.A. Gzelishvili supposed that "after smelting had been effected many times, only one side of the walls was dismantled, whence slag and then bloom was extracted."[30]

We do not possess the necessary archaeological material to determine the age of this workshop. Judging by the "thick" greyish patina covering the lumps of slag, we can confidently assign the workshop to the pre-Antique period. The data of archaeomagnetic investigation have proved that the last smelting in the "Djikhandjuri IV"

[28] Gzelishvili. *op. cit.*, pp-44-45, Fig. 12A, pl.I, 8.

[29] Ibid., Fig. 12 A,B, pl.I, 8.

[30] Gzelishvili. *op. cit.*, p. 45.

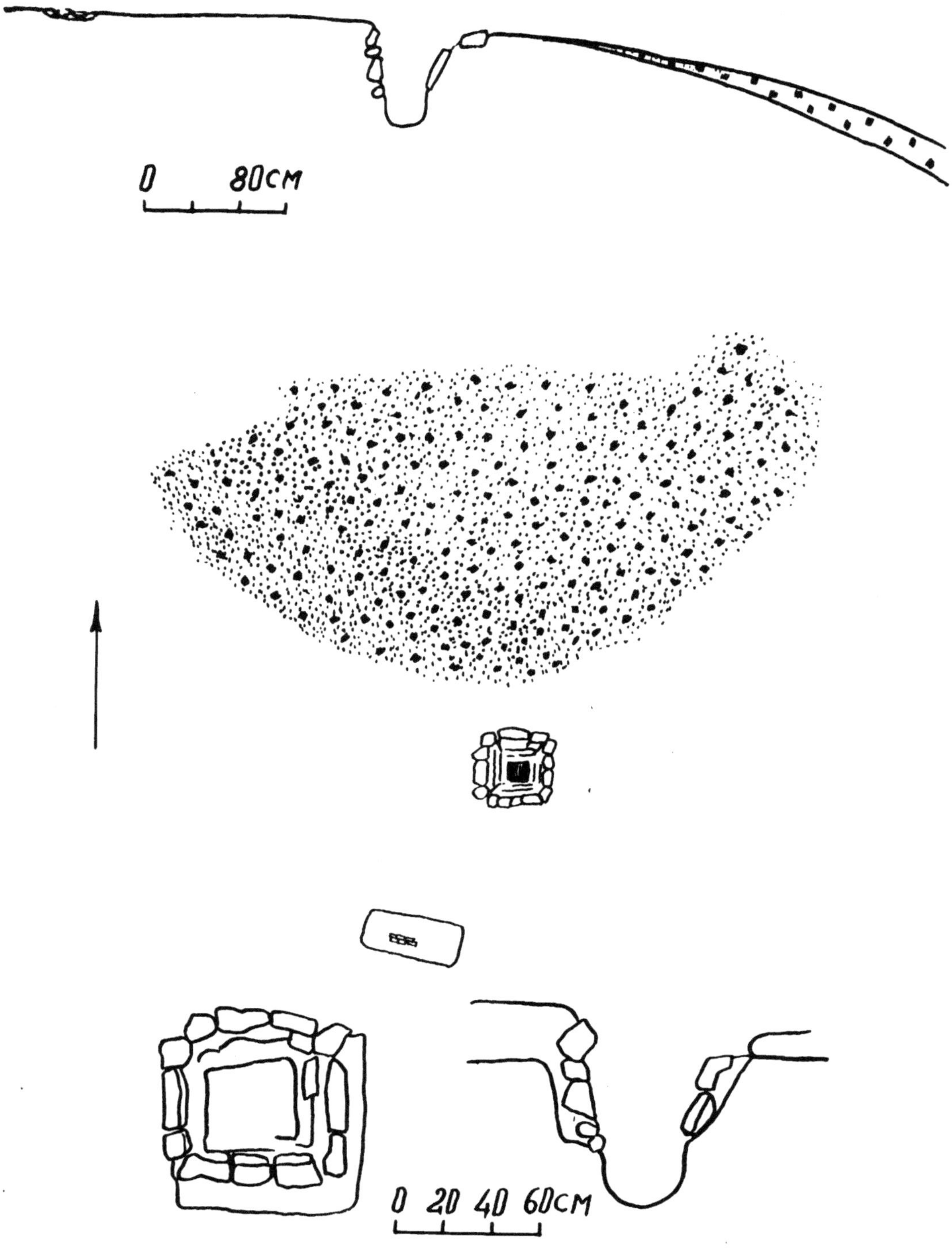

Fig. 11. Iron-smelting workshops "Djikhandjuri IV." General ground plan and sections.

furnace was done in approximately the middle of the ninth century B.C.[31]

5. IRON-SMELTING WORKSHOPS OF THE "TSETSKHLAURI" GROUP

The "Tsetskhlauri" iron-smelting workshops form a separate group in the Choloki-Ochkhamuri production area, designated, in specialist literature, by the name of "Tsetskhlauri." This group of workshops is scattered between the Choloki and Ochkhamuri rivers, separated by gullies, on the banks of small streams. They are mostly situated in the localities called Kapeti and Sapapria in the hydrosystem of the Khochicha, a right-bank tributary of the Ochkhamuri.

In the area over which the "Tsetskhlauri" workshops are scattered, the undersoils are clayey tufogenic strata of the Tertiary period, above which is a layer of diluvial yellow clay and yellowish-red loamy soil. At a depth of about

[31] Chelidze. *op. cit.*, p. 145.

three metres below the contemporary surface of the soil the laterites have a structure characteristic of clayey formations; from a depth of 3 to 20 or more metres alsolie clayey laterites, here and there presenting the structure of the bedrock; in these clayey strata the subsoil waters lie in most cases at a depth of 3-4 meters in macroporous soil. In the same area there are deposits of fire-clay of secondary origin, which, as mentioned above, is one of the principal conditions for the functioning of the bloomery furnace process of obtaining metal.

In the past, the territory between the Choloki and the Ochkhamuri, as we have stated above, was overgrown with virgin forests (wild chestnut-trees, oak, beech, cherry laurel, hornbeam, boxtree, rhododendron, etc.): splendid raw material for obtaining the necessary kind of fuel for smelting ores.

The above group of iron-smelting workshops consists of 15 units, of which, so far, only 3 have been excavated.

6. THE "TSETSKHLAURI I"[32] IRON-SMELTING WORKSHOP

The "Tsetskhlauri I" iron-smelting workshop was excavated by us in 1970. It has survived as follows: remnants of the iron-smelting furnace, the composite stone table (anvil) for hot forging of the spongy iron, and the waste dump (Fig. 12).

The remnants of the workshop were found on the land of the Tsetskhlauri state tea plantation, on the left edge of Sapatsria gully, Kapeti area, on the border of the plantation. The plot on which the remains of the workshop were found faces and slopes west.

The remnants of the workshop were overlaid by yellow diluvial clays 40-50 cm thick.

As it became evident during excavations, there were two iron-smelting workshops, one of which was destroyed in the 1930s when the soil was being prepared for tea planting. A few years later, during WWII, a military unit had built a temporary artillery position there which finally eliminated one of the workshops.

The "Tsetskhlauri I" (Fig. 12)[33] workshop has come down to us as a pit shaped like an upturned truncated pyramid. Judging by the "nests" in the upper two-thirds of the pit, this portion was faced with differently-shaped stones, while the one-third lower down was plastered with fire-clay. Neither the stone facing, nor the plastering have remained in situ. There is every reason to believe that they were ripped from the wall comparatively recently, in the 1930s. The walls and the bottom of the pit at a depth of 15-20 cm are burnt red by the intense heat.

[32] For a preliminary publication of the site, see: D.A. Khakhutaishvili. "Materials on the History of Ancient Georgian Iron Metallurgy (Ochkhamuri valley)." In: *Sites of South-Western Georgia*, Tbilisi, 1978, pp. 17-34. pls. III, IV, V, VI. (in Georgian, summary in Russian).

[33] The furnace was conserved *in situ*.

The interior of the furnace was filled in with yellow clay mixed with lumps of iron slag, fragments of the plastering and of nozzles of bellows, ashes and pieces of charcoal, as well as a few singed stones torn from the walls when the furnace was robbed.

Of the composite stone table nothing remains but a plot burnt red on which the table must have been laid, separate stones of which are scattered over the whole territory of the site.

The waste dump covered about 600 m^2 of space. The cultural level was up to 50 cm thick in places. Near the upper part of the dump, under a layer of ashes, a flat plot had been prepared for storing and "ripening" fire-clay.

The dump contained a great many lumps of iron slag (about 10 cubic metres) covered with a "thick" greyish patina, fragments of the clay plastering (about 4 cubic metres), nozzles of bellows, including one fragment with an orifice, and ceramic vessels, sherds of pots with zoomorphic handles and larger vessels decorated with broad fluting done by the "cutting" technique. Incidentally, fragments of a similar vessel were discovered when clearing the destroyed iron-smelting furnace (Figs. 12, 13). A large amount of ash and pieces of charcoal were recorded in the waste dump.

The lower edge of the dump reaches the bed of a nameless ravine with a stream that provided the workshops with water. To the south of the object, the macropores of lateritic Eocenic clays yield a plentiful source of fine potable water.

There are certain data that permit us to determine the age of the surviving iron-smelting workshop. In the first place, the ceramic vessels point to the beginning of the last millennium B.C. The pottery from "Tsetskhlauri I" (Fig. 14) finds analogues in strata III and IV of the Namcheduri settlement,[34] in the lower stratum of Tsikhisdziri and of some other settlements of the Colchian accumulation valley coastline.[35] Secondly, according to archaeomagnetic data, the last time "Tsetskhlauri I" was used for smelting was at about the beginning of the ninth century B.C. (declination = -2.0°, inclination = 75.0°, tension = 0.774 oersted).[36] Taking into consideration the fact that according to radiocarbon data the Namcheduri III and Tsikhisdziri IV settlements date back to the 10th century B.C., are justified in thinking that our site belongs to the early last millennium B.C. In any case, the workshop in question may be dated to a period no earlier than the tenth and no later than the eighth century B.C.[37]

[34] T.K. Mikeladze, D.A. Khakhutaishvili, *Namcheduri - an Ancient Colchian Settlement*. Tbilisi, 1985, pp. 14-16, pl. 24-51.

[35] Khakhutaishvili. *On the Chronology*., pp. 136-139.

[36] Chelidze. *op. cit.*, p. 145.

[37] Khakhutaishvili. *Materials on the History ...*, p. 27.

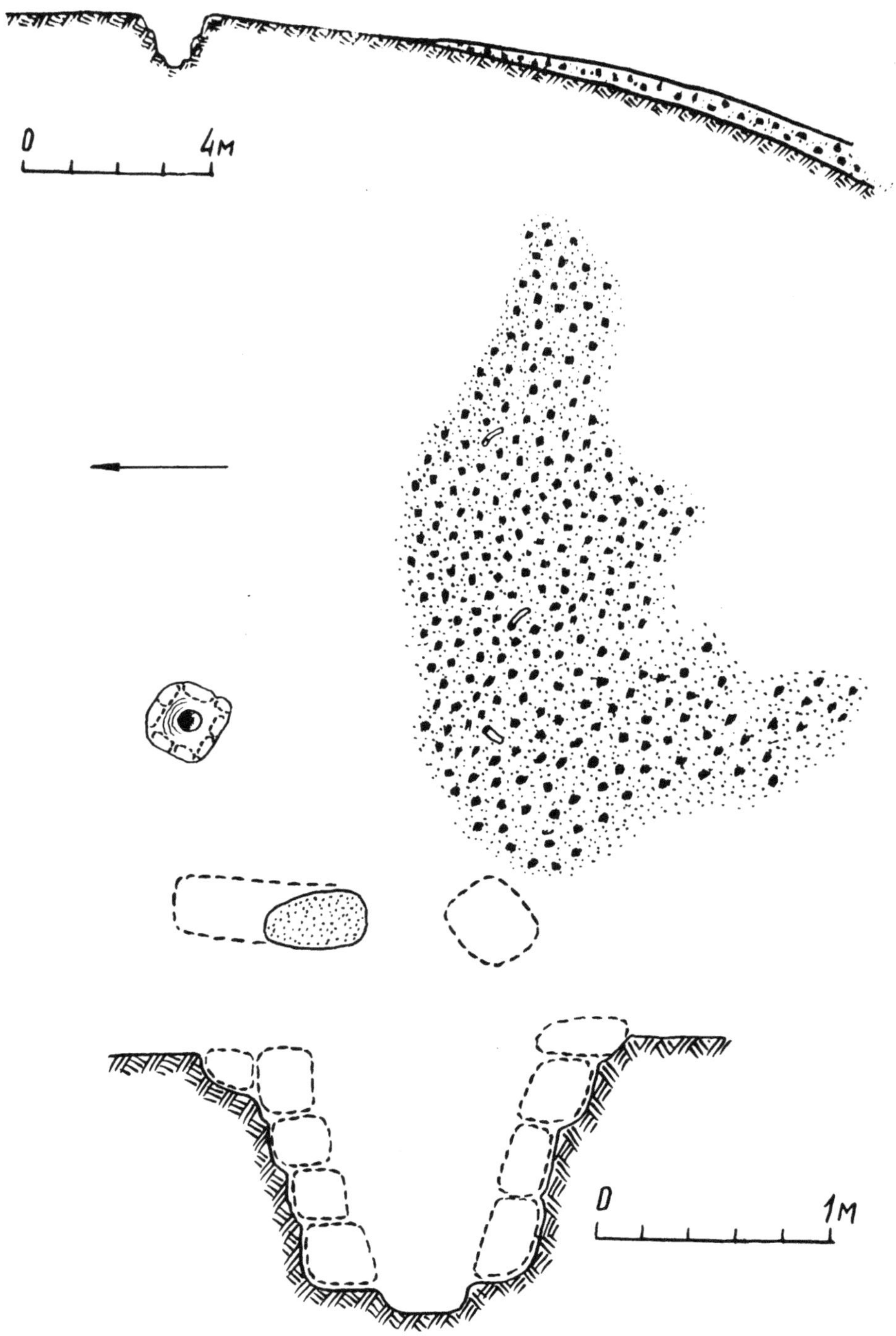

Fig. 12. Iron-smelting workshops "Tsetskhlauri I." General ground plan and sections.

7. THE IRON-SMELTING WORKSHOP "TSETSKHLAURI II"

The iron-smelting workshop "Tsetskhlauri II" (Fig. 15) was excavated by us in 1971.[38] The remnants of this workshop are situated to the north of the iron-smelting workshop "Tsetskhlauri I", at a distance of 1 km, in a neighbouring ravine also named Sapatsria. A little below the "Tsetskhlauri II" workshop the two Sapatsria gullies flow together forming one of the tributaries of the Khochicha brook.

Like all the previously described sites, this one was discovered owing to a slag heap on which there was no plant growth. The fact is that all the bloomery slags in Western Georgia of pre-Antique time contain a high percentage of ferrous oxide (from 13.05 to 48.15percent);[39] therefore, heated by the sun's rays, the slags burn the young shoots of plants.

[38] Khakhutaishvili. *op. cit.* pp. 27-33, pl.VII, VIII, IX, X.

[39] Gzelishvili. *Iron Manufacture...*, p. 82.

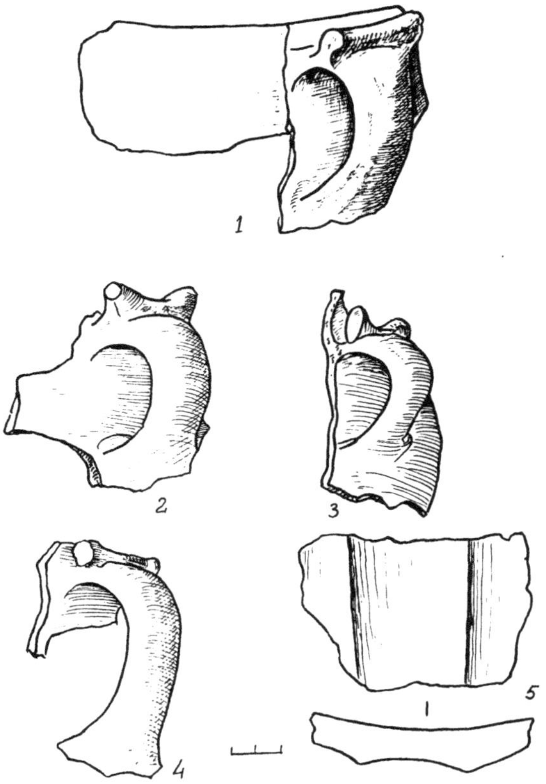

Fig. 13. "Tsetskhlauri I." 1-4 – fragments of zoomorphic handles of clay vessels; 5 – fragment of a thick-walled burnished vessel, ornamented with broad fluting.

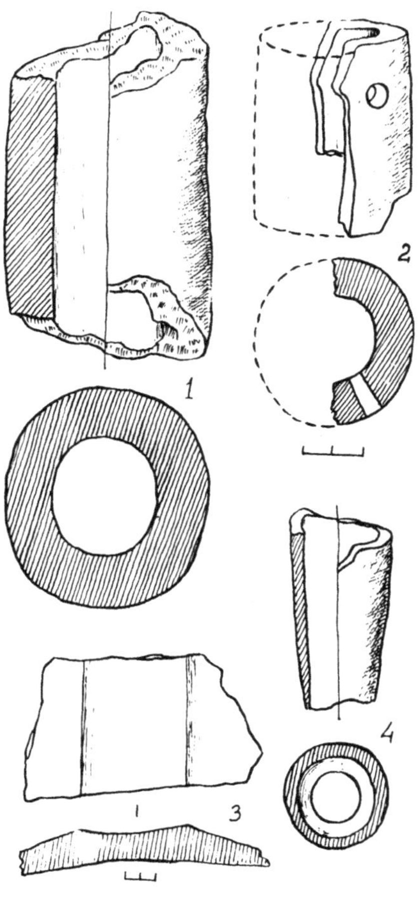

Fig. 14. "Tsetskhlauri I." 1,2,4 – fragments of flute-like clay nozzles for bellows; 3 – fragment of a thick-walled burnished vessel, ornamented with broad fluting.

The plot where the remnants of the "Tsetskhlauri II" iron-smelting workshop were identified is on a slope and is exposed to the north; here, close to a gully and the stream running down it a small platform was made for the workshop of which there remains the iron-smelting furnace, and some part of the table (anvil). All the territory of the workshop except the waste dump is occupied by a tea plantation. The overall surface of the excavation is 400 m^2; the soil is exclusively that which is widespread in all the subtropical foothills of the Eastern Black Sea littoral, yellow clay of diluvial origin.[40]

The iron-smelting furnace "Tsetskhlauri II"[41] has survived as a pit shaped like an upturned truncated pyramid. The traces of the furnace were covered by a layer of worked yellow clay 30-35 cm thick. The lower part of the furnace ended in a spherical depression 40 cm in height, plastered with a thick layer of fire-clay. The upper part of the furnace was faced with stone (Fig. 15); at a height of 40 cm above the bottom, the master smelter had left out a ledge, 20-25 cm wide, for the stone masonry to rest on. The slabs of the facing were purplish on the surface and reddish in fractures some of which had melted. The stone facing and the clay plastering of the furnace were in good condition.

The dimensions of the furnace were as follows: height 100 cm; width at top 60 cm, in the middle 45 cm, at the bottom 10-15 cm.

The inner part of the furnace was filled in with clay in which were occasional lumps of iron slag and charcoal, fragments of the clay plastering and of bellow nozzles; also ashen and burned stones, the presence of which leads us to think (seeing that the stone facing of the furnace was well preserved) that the part of the furnace above-ground was also built of stone with clay. The volume of the stones found inside the furnace was not large, about one third of the overall height.

At a depth of 15-20 cm, the soil around the furnace was burnt red by the heat.

Of the table (anvil) there remained actually only a flat area burnt red and a few scattered stones of the pavement. The waste dump lay to the north and north-west of the workshop and covered a large area of about 300 m^2, sloping towards the stream (gully). The cultural stratum was sporadically 40-50 cm thick. In the north-western part of the dump, in an area of about 24 m^2 there was a stock of fire-clay of about 7 m^3 in volume, stored here for "ripening."

The dump contained a great many lumps of iron slag (about 10 m^3) covered with a "thick" greyish patina, fragments of ash-coloured clay plastering of the furnace (more than 3 m^3), sherds of pottery and fragments of nozzles for bellows; also ash, fragments of charcoal and scorched stones.

Burnished household pottery is represented by fragments of zoomorphic handles, flat bases, as well as by larger vessels ornamented with broad and narrow fluting (Figs. 16, 17). The ceramic ware made of pure well-sifted clay with an admixture of fine-grained sand is characterized by its colour: fired an even grey, black or dark-brown. It was made on a fast potter's wheel at the beginning of the first millennium B.C. and is of a kind widespread throughout seaboard Colchis.

The pottery dating from the ninth century B.C. is a guide to determining the age of the site in question, so is the comparative chronology of the workshops of the "Tsetskhlauri" group that was dated by the archaeomagnetic method.[42] We should immediately mention that we had dated "Tsetskhlauri I" and "Tsetskhlauri II" by archaeological data and suggested that they be used as basic sites for archaeomagnetic determination.

The Colchian iron-smelting furnaces are fine material for archaeomagnetic study and for dating sites, but we are, as yet, only on the way to establishing the basic curve of the changes in the parameters of elements of the ancient geomagnetic field in the Colchian region; the date above may serve to establish a comparative chronology.[43] The two iron-smelting workshops described above have very close affinities with Namcheduri settlement III, with the lower strata of Tsikhisdziri and with a number of sites contemporary with them.[44] According to archaeological and radiocarbon analysis data, they all belong to the 10th-ninth centuries B.C.[45]

Thus, the iron-smelting workshops "Tsetskhlauri I" and "Tsetskhlauri II" belong to different parts of the ninth century B.C.

8. IRON-SMELTING WORKSHOP "TSETSKHLAURI III"

The "Tsetskhlauri III" iron-smelting workshop was excavated in 1974. It was identified in an area called Bogili on a tea plantation belonging to the local state farm, at the top of a hill, part of a watershed between the Choloki and Ochkhamuri rivers.

It was impossible to find the furnace. It must have been destroyed when the tea-bushes were planted, which is also suggested by numerous scorched stones scattered

[40] See preliminary publication of the sites by D.A. Khakhutaishvili. "Archaeological Excavations of an Ancient Colchian Centre of Iron Metallurgy in the Choloki Valley in 1971" (in Georgian). *Matsné*, 1974, No. 4, pp. 199-202.

[41] Furnace conserved *in situ*.

[42] D.A. Khakhutaishvili. *On the Chronology*... p. 139. Z.A. Chelidze. *Results*... pp. 142-145.

[43] Ibid., p. 44

[44] Khakhutaishvili. *On the Chronology*., p. 135 ff.

[45] T.K. Mikeladze, D.A. Khakhutaishvili. *An Ancient Colchian Settlement*. pp. 14-15.

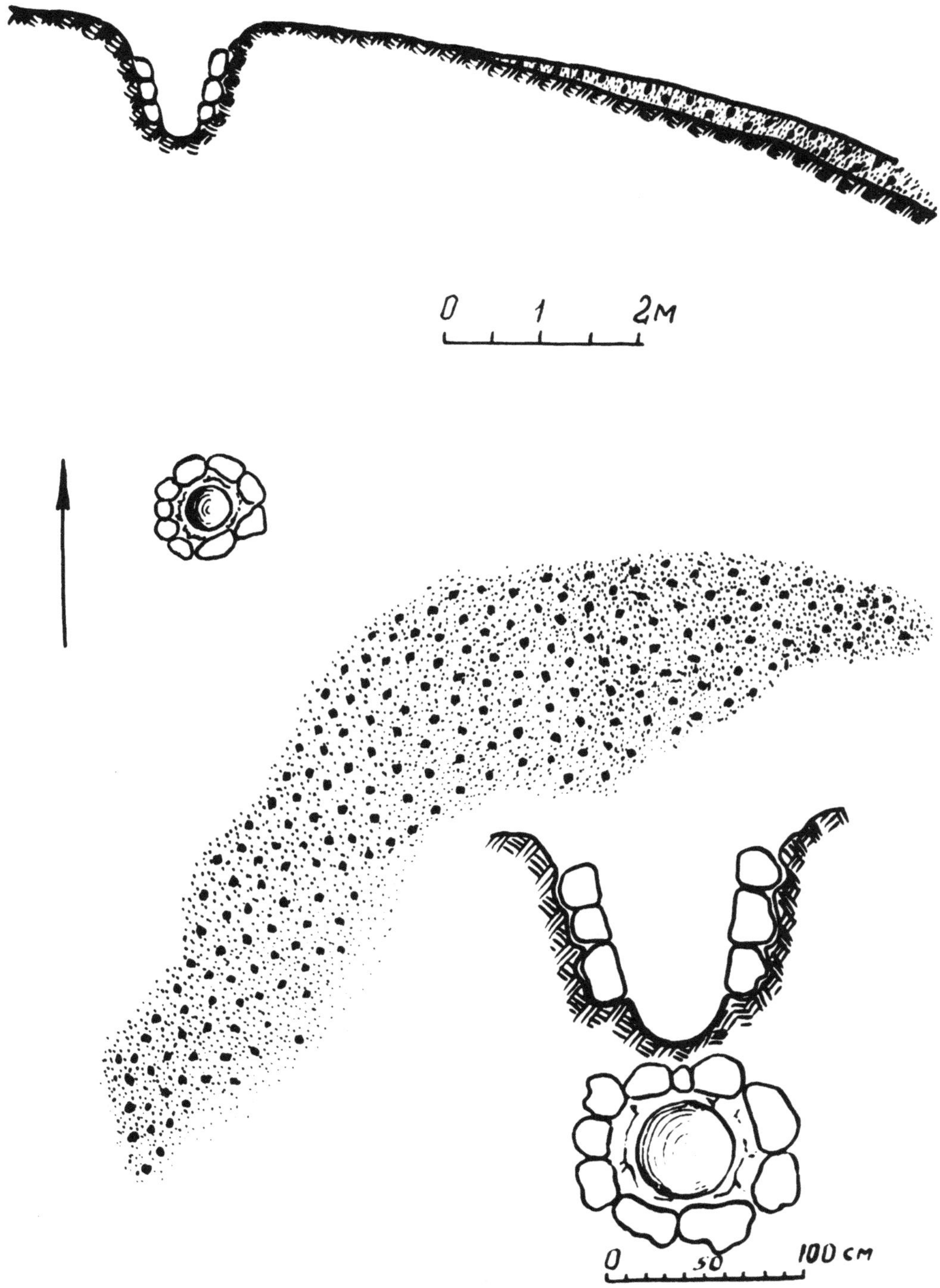

Fig. 15. Iron-smelting workshops "Tsetskhlauri II." General ground plan and sections.

over the slopes of the hill exposed to the north-west. The composite table (anvil) has not survived either.

The waste dump lay on the north-western slope of the hill and covered about 200 m^2 The dump, 40 cm deep in places, yielded a great many lumps of iron slag (totalling about 6 m^3) covered with a "thick" greyish patina, fragments of fire-clay plastering of the furnace, of nozzles for bellows, a considerable number of scorched stones, ashes, bits of charcoal, and, principally, many pottery sherds. The most noteworthy of these are fragments of a thick-walled vessel fired red and decorated with white fluting done by cutting.

Such potsherds occurred among the materials of the iron-smelting workshops "Tsetskhlauri I", "Tsetskhlauri II", "Charnali I", "Charnali II", of the ancient settlements "Namcheduri III", "Tsikhisdziri IV", "Kulevi IV" and elsewhere.[46] All these sites fit in with the archaeological

[46] D.A. Khakhutaishvili. *Materials on the History of Ancient Colchian Iron Metallurgy (Ochkhamuri Valley)*, pp. 33-34.

Fig. 16. "Tsetskhlauri II." 1,2,4,6 – fragments of zoomorphic handles and walls of clay vessels, ornamented with narrow fluting; 7,8 – fragments of thick-walled burnished vessels, ornamented with broad fluting.

frame from the very end of the second millennium B.C. to the eighth century B.C. inclusively.[47]

A number of problems connected with this group of iron-smelting workshops will, most probably, be solved by future excavations planned for the next few years.

[47] D.A. Khakhutaishvili. *On the Chronology...*, pp. 135-141.

9. IRON-SMELTING WORKSHOPS "LEĞVA I"

The iron-smelting workshops "Leğva I" (Fig. 18) were unearthed by I.A. Gzelishvili in 1961[48] on the tea plantation of Leğva village collective farm, to the left of the Kobuleti-Ozurgeti highway, on a smallish promontory-like hill looming over the road by about 10 metres.

[48] I.A. Gzelishvili. *Iron-Smelting in Ancient Georgia*, pp. 45-47, Figs. 13A, 13B, pl. II, 2-3, IXB, 1-5, 10.

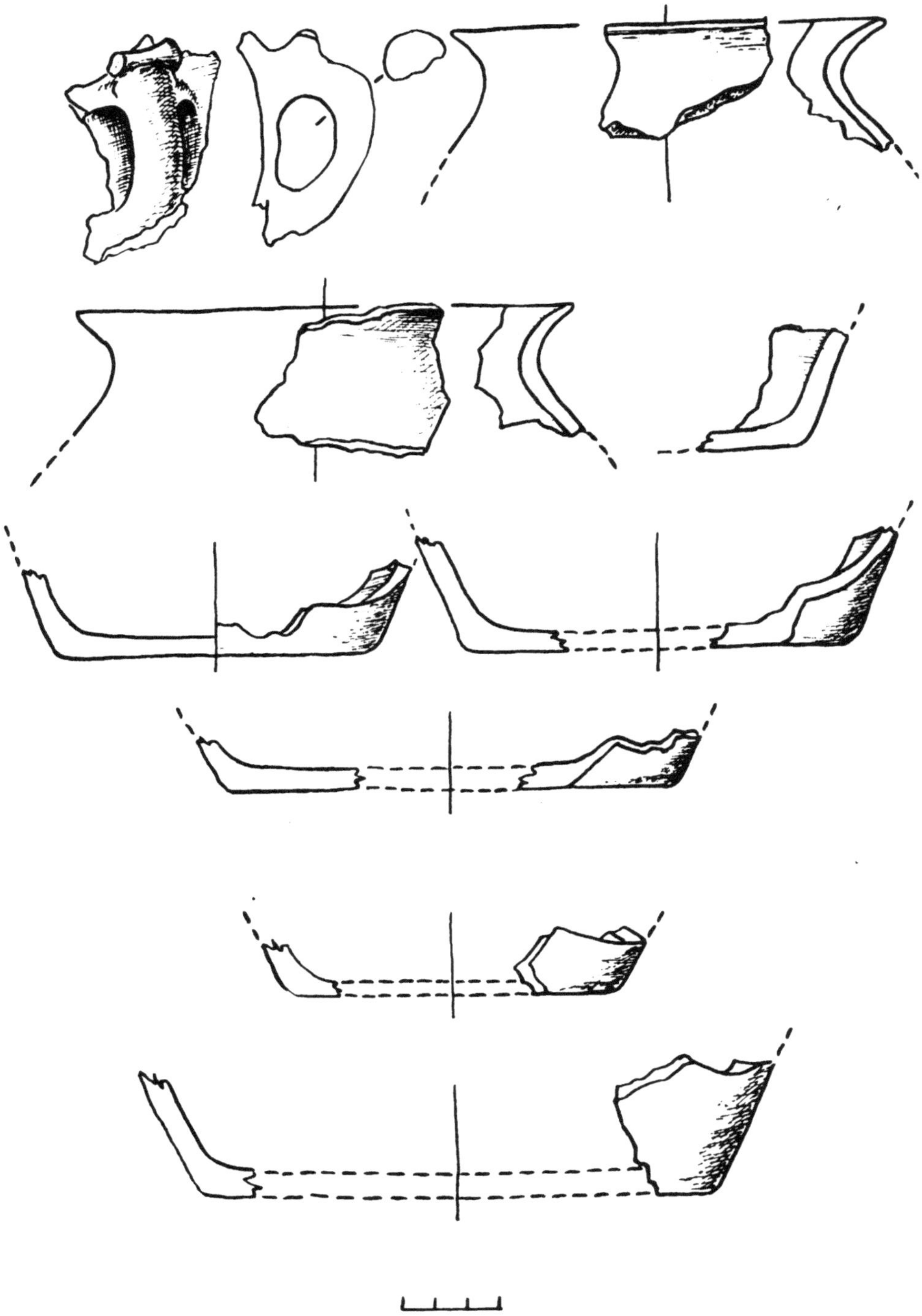

Fig. 17. "Tsetskhlauri II." 1 – fragment of zoomorphic handle of clay vessel; 2,3 – fragments of clay pots; 4,9 – fragments of bases of the vessels.

The hill with its southward exposure "is composed of clayey tufogenic Tertiary rock covered by a thinnish layer of yellow diluvial clays and yellowish-red loams. In the western part of the hill a porphyritic dyke breaks in a narrow strip through a denudation of sedimentary strata."[49]The site was identified thanks to the waste dumps on the south-eastern and south-western slopes of the promontory-like hill. For reasons mentioned above, the central parts of the dumps were bare of plant growth.

On the top of the hill was a smallish platform (20×20 metres) from which the expedition removed a sterile layer to a depth of 50 cm, which revealed, in the centre of the excavation, the outlines of the upper horizontal section of two iron-smelting furnaces and patches of diluvial clays burnt red, most probably the bases of the composite tables for forging the bloom.

Thus, "Leğva III" revealed the traces of two iron-smelting workshops which belong to the peripheral group

[49] Gzelishvili. *op. cit.*, p. 46.

of the Choloki-Ochkhamuri centre of the Colchian ancient iron-metallurgical area.

a. The Iron-Smelting Workshop "Leğva I, 1"

The iron-smelting workshop "Leğva I, 1" has survived in the form of remnants of the iron-smelting furnace, the base of the composite table (anvil) and the waste dump.

Of the iron-smelting furnace "Leğva I, 1", (Fig. 18)[50] only the subterranean part remains. It is in a pit shaped like an upturned truncated pyramid, the lower part of which is plastered with a thick layer of fire-clay, and the upper part is lined with various roughly hewn stones. The high temperature had melted the surface of the furnace here and there, at the same time lending it a purplish colour.

The inner space of the furnace was filled with clay mixed with scorched stones, lumps of iron slag, bits of charcoal, ashes, fragments of clay plastering, ash-grey in colour, as well as nozzles for bellows; in places were found traces of charcoal dust.

Of the table (anvil) there remains only a patch of diluvial clay burnt red, 200×100 cm.

The waste dump of the "Leğva I, 1" furnace was mostly on the south-eastern slope of the hill (Fig. 18) and covered approximately 100 m^2 of ground. The dump was up to 50 cm deep in places. It contained a great many lumps of iron slag (about 5 cubic metres) covered with a "thick" greyish patina, fragments of the clay plastering of the furnace, fragments of nozzles for bellows, ashes, charcoal, dust, etc.

There are certain indirect data to determine the age of "Leğva I, 1": a) according to its structure and the patina on the slag, it belongs to the pre-Antique period; b) archaeomagnetic data assign the last smelting in the furnace to the end of the tenth century B.C. (declination = -20°, inclination = 72°, tension = 0.426 oersted).[51] There are no data to contradict this.

Thus, the iron-smelting workshops "Leğva I, 1" may be assigned to the 10th-ninth centuries B.C.[52] Here we might mention that Gzelishvili considered both furnaces of the "Leğva I, 1" site to belong to one workshop, and, on the basis of material yielded by the "Leğva I, 2" furnace when it was cleared, he assigns the "Leğva I, 1" furnaces to the 12th century B.C.[53] It turned out in the process of further investigations that such a dating was unfounded,[54] the error being 200 or 300 years, to which, incidentally, the data of "Leğva I, 2" point to.

b. Iron-Smelting Workshop "Leğva I, 2"

The iron-smelting workshop "Leğva I, 2" consists of the iron-smelting furnace, the base of the composite table (anvil) and the waste dump. It is situated on the same plot as the "Leğva I, 1" workshop (Fig. 18).

The "Leğva I, 2" iron-smelting furnace is only 100 cm east, away from the "Leğva I, 1." It is a pit shaped like an upturned truncated pyramid, the upper part faced with rough-hewn stones, in the lower part with a thick layer of fire-clay. The stones are melted down here and there and have acquired a purplish colour.

The table (anvil) has survived as a patch of diluvial clay burned red, dimensions 200×100 cm. The scorched stones of the composite table were scattered over the entire territory of the excavation and down the hillsides.

The waste dump was on the western slope of the hill where we collected lumps of slag totalling about 6 cubic metres, about 2 cubic metres of fragments of ash-grey clay plastering, fragments of bellow nozzles, scorched stones from the structure of the furnace and the composite table, ash, pieces of charcoal, a stock of fire-clay, shapeless potsherds, etc.

For dating the above shop we had at our disposal a number of ceramic sherds (Fig. 19, 4-9; Fig. 20) the chronological limits of which lie between the 11th and eighth centuries B.C. Seeing that the pottery from "Leğva I, 1" is more archaic and has affinities with the ancient settlement "Tsikhisdziri IV", the workshops "Leğva I, 2" may be assigned to the 11th-10th centuries B.C. In future, when our physicists have mastered the methods of dating slag from bloomery furnaces, this dating may be corrected, as well an the age of some other sites, and the exact dates of all the sites collected in these excavations and connected with iron manufacture will be recalibrated.

As mentioned above, the Choloki-Ochkhamuri centre of ancient iron manufacture lies in the area of the middle reaches of the Choloki and Ochkhamuri rivers; at their confluence, in a locality named Pichvnari-Choloki and Ispani, on an accumulation seaboard valley north of the Kobuleti seaside resort, is a group of exceedingly important archaeological sites belonging to various periods, beginning with the early Bronze Age (a settlement covered by a stratum of peat, at Ispani) and ending with the Hellenistic period.[55] In connection with the study of the Choloki-Ochkhamuri iron producing centre, especial attention should be paid to a seaboard settlement of the New Black Sea terrace whose investigation was launched by N.V. Khoshtaria (1953, 1959) and carried on by A.T. Ramishvili (1960-1964, 1967).[56]

---

[50] The furnace was conserved in situ

[51] Chelidze. *op. cit.*, p. 145.

[52] Cf. Khakhutaishvili, *Materials on the Story of Ancient Georgian Metallurgy*, pp. 18-20.

[53] Gzelishvili. *Iron-Smelting in Ancient Georgia*, p. 51.

[54] Khakhutaishvili. *On the Chronology…*, p. 140, ff.

[55] Khakhutaishvili. The Batumi Research Institute and the Archaeology of Georgia, pp. 54-62.

[56] Ramishvili. "Archaeological Excavations on Seaside Stations in Pichvnari (Kobuleti) in 1960-1964 and 1967". *SAS* 1975. Idem, "On the

N.V. Khoshtaria, I.A. Gzelishvili, D.A. Khakhutaishvili, A.T. Ramishvili and others have stated that master smelters working at the seaside used magnetitic sand as raw material. Later on, A.T. Ramishvili specially dwelt on this problem and made a not unsuccessful attempt, presenting concrete material to substantiate his assertion that the seaside stations in the Eastern Black Sea littoral engaged mainly in obtaining raw material for iron manufacture. It is only fair to mention that other authors had made this hypothesis before I.A. Ramishvili tried to corroborate it by bringing forward concrete evidence.[57]

At present, a considerable amount of material has been collected testifying to the fact that the prosperity of the seaboard settlements was largely dependent on the growth of iron production. Archaeological material from Ureki, Pichvnari and Bobokvati enable us to state that the so-called permanent and temporary settlements on the New Black Sea terrace appeared at the developed stage of the Phanagorean maritime regression, contemporaneous with the initial stage of economic mastery of iron in Colchis. The production of iron in bloomery furnaces was initiated there several centuries before seashore settlements arose on the New Black Sea terrace, the formation of which began, incidentally, some time in the latter half of the third millennium B.C., in the period of the New Black Sea transgression, and ended about the middle of the second millennium B.C.[58] On this basis we may surmise that the emergence of seaside settlements principally engaged in obtaining raw materials for iron manufacture was conditioned by the growing demand for iron and its wide utilization at the end of the second and at the beginning of the first millennium B.C.

Purpose of Stations with 'Textile' Ceramics on the Eastern Black Sea Littoral." *Soviet Archaeology*, 1975, 4 pp. 36-44; idem, "On the Chronology of the Eastern Black Sea littoral Stations at Pichvnari." *Matsné*. 1974, 2, pp. 103-111.

[57] Ramishvili. "On the Purpose of Stations with 'Textile' Ceramics on the Eastern Black Sea Littoral," pp. 36-44. To answer A.I. Ramishvili's statement that "neither does D.A. Khakhutaishvili, who had devoted special papers to problems of ancient iron metallurgy in Colchis, mention seaside stations" (p. 40), I am obliged to quote some excerpts from my paper, "On the Sources of Colchian Iron Metallurgy" (1964) to which I.A. Ramishvili refers: "It turns out that iron smelting from magnetitic sand was proper not only to Chalybes (Chaldeans), but also to communities of iron-makers from other regions of historic Colchis. Moreover, we have grounds to believe that the use of magnetitic sands as raw material was more natural in the Choloki river basin than in Chalybia (Chaldea). The iron manufacture centre in the Choloki basin situated in the upper reaches of the river received raw material from the seashore. "If we take into consideration that this centre comprised several dozen workshops scattered over large territory, we have to admit that this centre was in close contact with seaboard settlements". (D.A. Khakhutaishvili. *op. cit.*, pp. 52-53). Mentioning several other data, we come to the conclusion that "there existed a certain economic unity between the iron manufacturing centre and the seaside settlements; ibid p. 55). The same may be said about Gzelishvili's approach to this question when he started excavations of seaside settlements at Pichvnari in the same year, 1960, and later on entrusted this work to A.I. Ramishvili. The fact that I.A. Gzelishvili turned his attention to seaside settlements in connection with studying ancient Colchian iron metallurgy speaks for itself.

[58] Ch.P.Djanelidze. *Palaeography of Georgia in the Holocene*. Tbilisi, 1980, pp. 45-52 D.A. Khakhutaishvili. *Nature and Man in the Colchian Littoral in the Holocene*.

The fact that such settlements sprang up in greater numbers on those parts of the New Black Sea terrace that were close to centres of ancient Colchian metallurgy once more corroborates the earlier suggested idea that the iron-producing centres on the Colchian seaboard and the seaside settlements on the New Black Sea terrace rich in magnetitic sands were a single economic entity, a single production complex.[59]

The Choloki-Ochkhamuri mining centre most probably promoted the growth of the settlement at the confluence of the Choloki and Ochkhamuri rivers, and caused Greek trading ships to cast anchor there, and then a Greek community to take root there and engage in trading iron and possibly in the manufacture of steel objects.

At any rate, at first a large permanent settlement arose on the seashore (15th-7th centuries B.C.),[60] later which grew into a town (from the fifth century B.C.) where a Greek and indigenous Colchian communities lived and worked in peace; to a great extent, this was favoured by its existence as a considerable metallurgical centre.

We are inclined to think that this considerable iron production centre presupposes sufficiently well-organized labour; all the more that the producing workshops and the sources of raw material were at a distance of a few kilometres from each other.

We should note here that we have not as yet succeeded in discovering contemporary settlements in the region where the ancient production centre functioned. If this is corroborated in future, it will be necessary to look for where the master-metallurgists lived and when they actually engaged in iron smelting. Most probably, iron smelting was seasonal work in ancient Colchis, when the population was free from crop cultivation, i.e. in winter and in early spring, as has been found in many regions of the ancient world.

Thus, the iron manufacturing centre we have investigated, with its more than one hundred workshops identified about the middle courses of the Choloki and Ochkhamuri rivers, as well as a contemporary group of sites in their lower reaches, appear to be two parts of one production centre (of course, as it was understood then). Only such an approach may explain the situation that prevailed in the southern part of the Colchian valley.

[59] D.A. Khakhutaishvili. *On the Sources of Ancient Colchian Iron Metallurgy*, p. 55; idem. "Raw Materials Resources and the Techniques of Iron Manufacture in Ancient Colchis." In: *History of Mining and Metallurgy*, Tbilisi, 1979, pp. 86-88.

[60] T.K. Mikeladze, D.A. Khakhutaishvili. *Namcheduri, an Ancient Colchian Settlement*. Tbilisi, 1985.

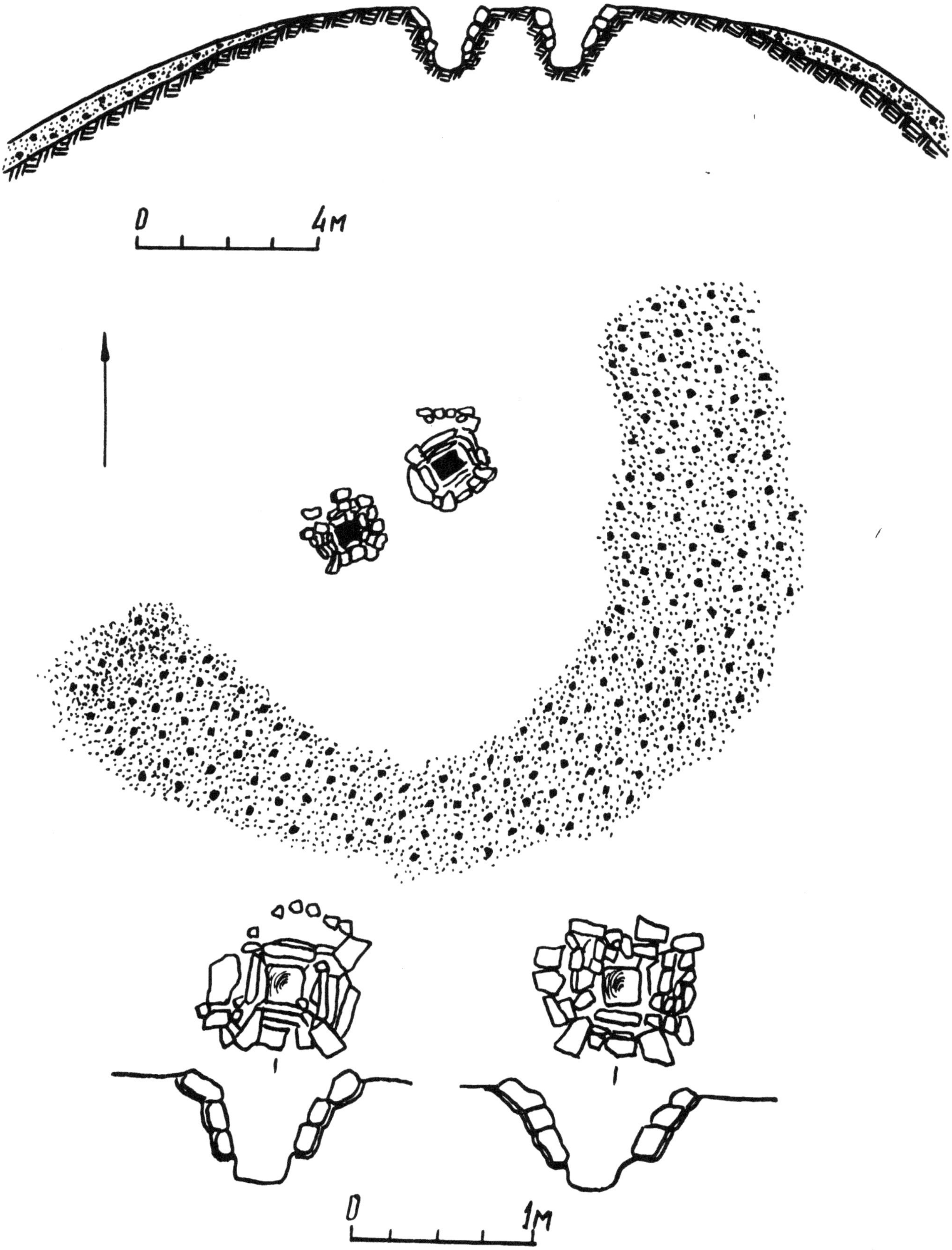

Fig. 18. Iron-smelting workshops "Leğva." General ground plan and sections.

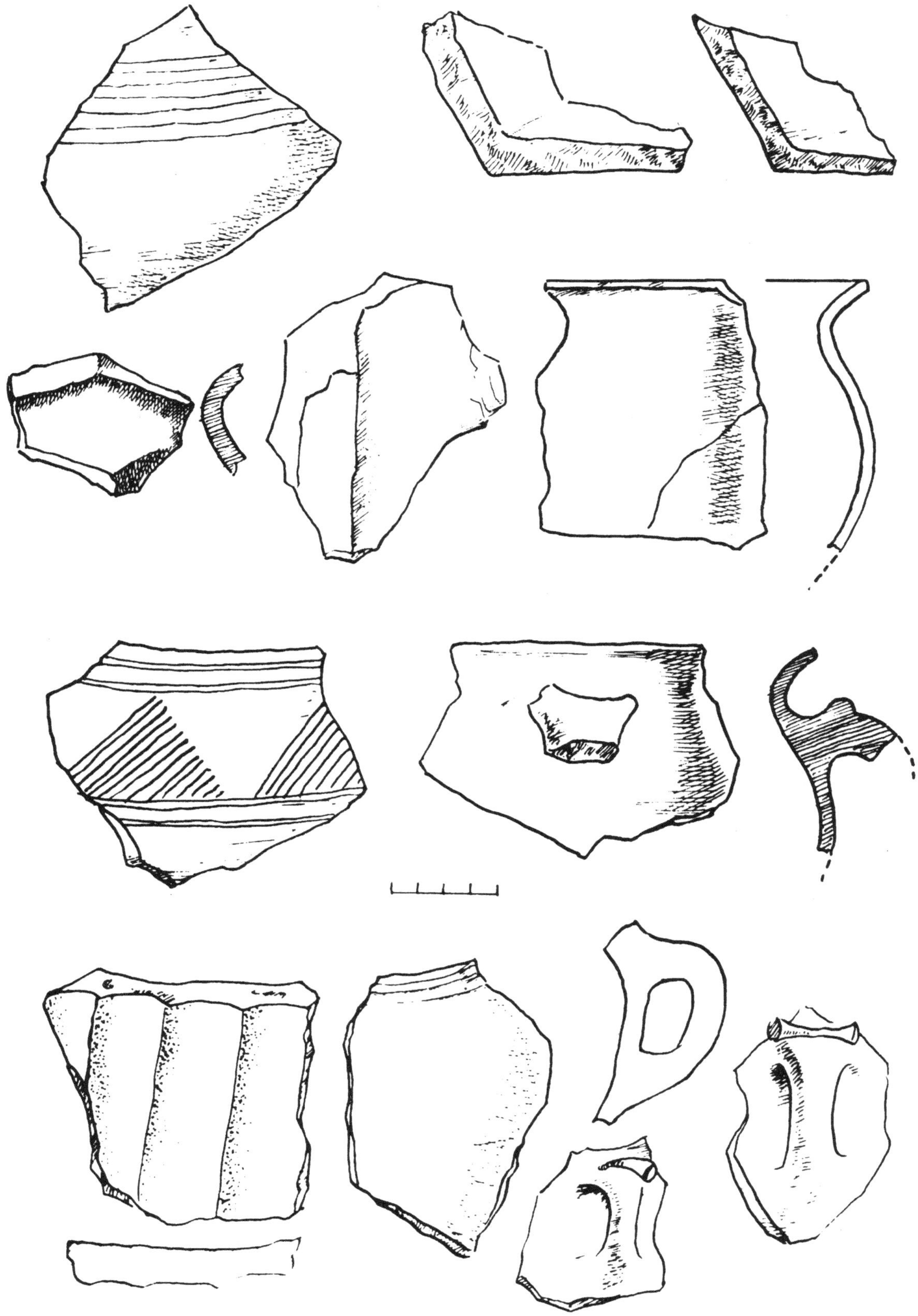

Fig. 19. Fragments of clay vessels from various iron-smelting workshops: 1-3 – from "Djikhandjuri I"; 4-8 – from "Leğva"; 9 – from "Charnali"; 10-11 – from "Djikhandjuri III"; 12-13 – from "Djikhandjuri II."

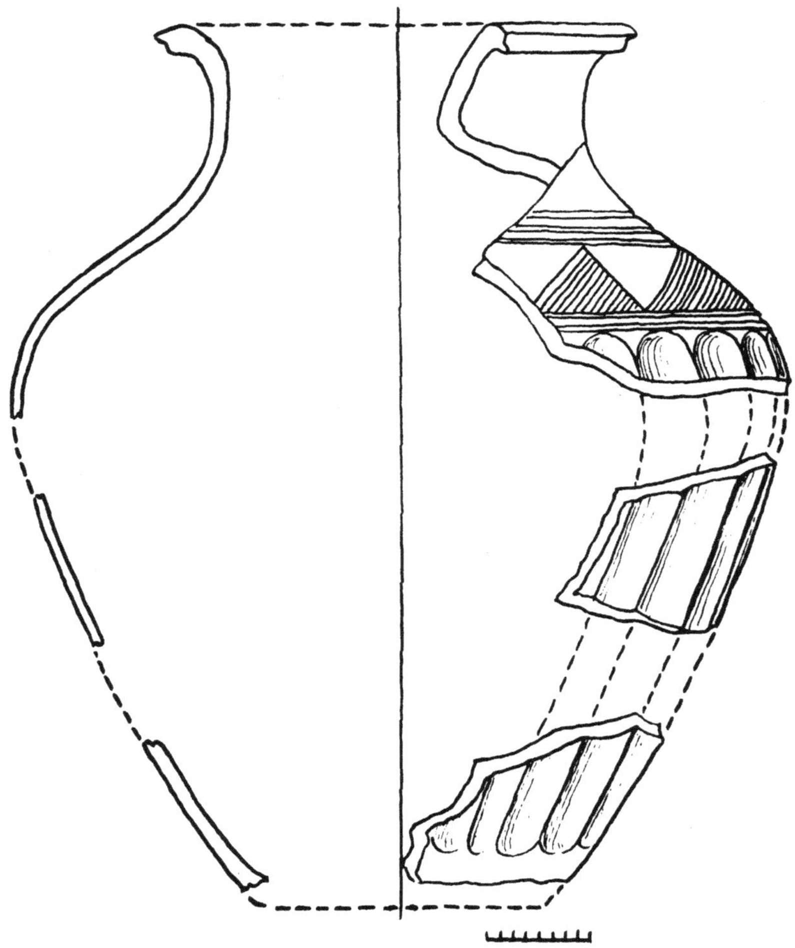

Fig. 20. Vessel from "Leğva I" restored.

CHAPTER III

THE SUPSA-GUBAZEULI IRON-SMELTING CENTRE

The Supsa-Gubazeuli iron-smelting centre, the most considerable of all Colchian sites of this group, is situated in the region around the middle course of the Supsa river, in the valley of its left tributary, the Gubazeuli, at a distance of 30-40 km from the Choloki-Ochkhamuri manufacturing centre (Fig. 20A). The central part of this area is on the lands of the villages Askana, Mziani (Pichkhis Djvari), Dabali Etseri, Mshvidobauri and Nagomari, while the peripheral workshops are scattered over the lands of the villages Gora-Berezhouli, Kvenobani, Khidistavi etc. This centre is different from others on several points:

a. It includes more sites of ancient iron manufacture than any other centre;
b. It continued functioning from the Late Middle Bronze and Early Late Bronze down to the Early Antique period inclusively;
c. It was close to a more ancient centre of copper-bronze manufacture, the traditions of which forwarded the inception of iron manufacture;
d. Not far from the very centre (10-15 km away) lay the Vakidjvari-Korisbudé ore-bearing massif and ample resources of magnetitic sands at Ureki.

The centre of this iron producing area lay in the valleys and depressions in the basin of the Gubazeuli river and its tributaries (the Gulepa, Kurepa, Tskaltsitela I, Tskaltsitela II) mainly in the area between the Bakhvistskali and the Kurepa.

The soil in this region consists mostly of large deposits of laterites formed in a subtropic zone by the erosion of various Tertiary tufas. They are overlaid by yellow diluvial clay containing a considerable amount of magnetite grains. These grains are washed down into the Supsa River which falls into the Black Sea near Ureki. That is where the richest deposits of magnetitic sands[1] lie all along the shore; these deposits, the richest on the entire Eastern Black Sea littoral, are one of the main factors favouring the production of iron.[2] Here are also deposits of fire-clay and refractory clays, as well as forest massifs characteristic of Colchis.

One group of workshops belonging to the centre under consideration but territorially away from it and receiving another kind of raw material from another source deserves special mention. These workshops are situated in the upper reaches of the Natanebi river, at the foot of the Bakhmaro resort, in a place named Korisbudé in a zone where polymetallic ores are obtained. This group of workshops consisting of about 10 sites showing traces of metal production, is 10-15 km away from villages or settlements and cannot be reached by car; for this reason it is difficult to excavate there.

It should be especially noted that at present the foothill zone between the Supsa and the Choloki rivers is the richest site of ancient iron metallurgy known in the region of the Caucasus and the Near East.

1. IRON-SMELTING WORKSHOP "ASKANA I" (excavated in 1974)

The remnants of the "Askana I" iron-smelting workshop were identified in 1974 on the tea plantation of Askana village collective farm (team No.7), to the left of the Tskaltsitela I ("Red Water") brook, 200 metres west of the former holding of Meliton Ninidze, on the northern slope of a nameless ravine, near an automobile road to the plantation. Visually, there was only one indication, a largish heap of slag lumps and plaster of the furnace, free of any vegetation. Digging exposed some badly damaged remains of an iron-smelting furnace and the base of a composite table (Fig. 21).

There were actually only traces of the "Askana I" iron-smelting furnace. It was destroyed when a trench was being dug. Only the west wall of the furnace remained and a burnt strip of reddish-yellow clay. Lumps of slag and pieces of bellow nozzles were picked out of the pit.

The base of the composite table (anvil) was found 250 cm north of the pit, actually an area 200x70 cm, burnt red (Fig. 21).

The waste dump, as we have mentioned, occupied a south-oriented slope and covered over 100 square metres. It contained a large number of lumps of slag with a thick greyish patina, greyish-ash coloured fragments of the clay plaster of the furnace, a stone hammer (of diorite), a stone mortar for pounding flux and ore, fragments of cone-shaped clay nozzles for bellows (Fig. 22), a number of potsherds: handles, bases, rims of pots including thick-walled vessels decorated outside with broad fluting executed by the "cutting" technique (Fig. 23). As is known, a hammer of stone (diorite) with grooves for

[1] *The Mineral Resources of the Georgian SSR*. Tbilisi, 1935, pp. 244-245. *The Natural Resources of the Georgian SSR*, I, Moscow, 1958. p. 83.

[2] The first information about dumps of iron slag on the territory of Askana village, Ozurgeti district, came from D.G. Megreladze, senior researcher at the I.A. Djavakhishvili Institute of History, Archaeology and Ethnography, Academy of Sciences of the Georgian SSR. Historical and geographical exploration corroborated the presence of pre-Antique slag there. It was only in 1974 that we were able to make a planned study of the site and to determine that we were dealing with a large iron-producing centre.

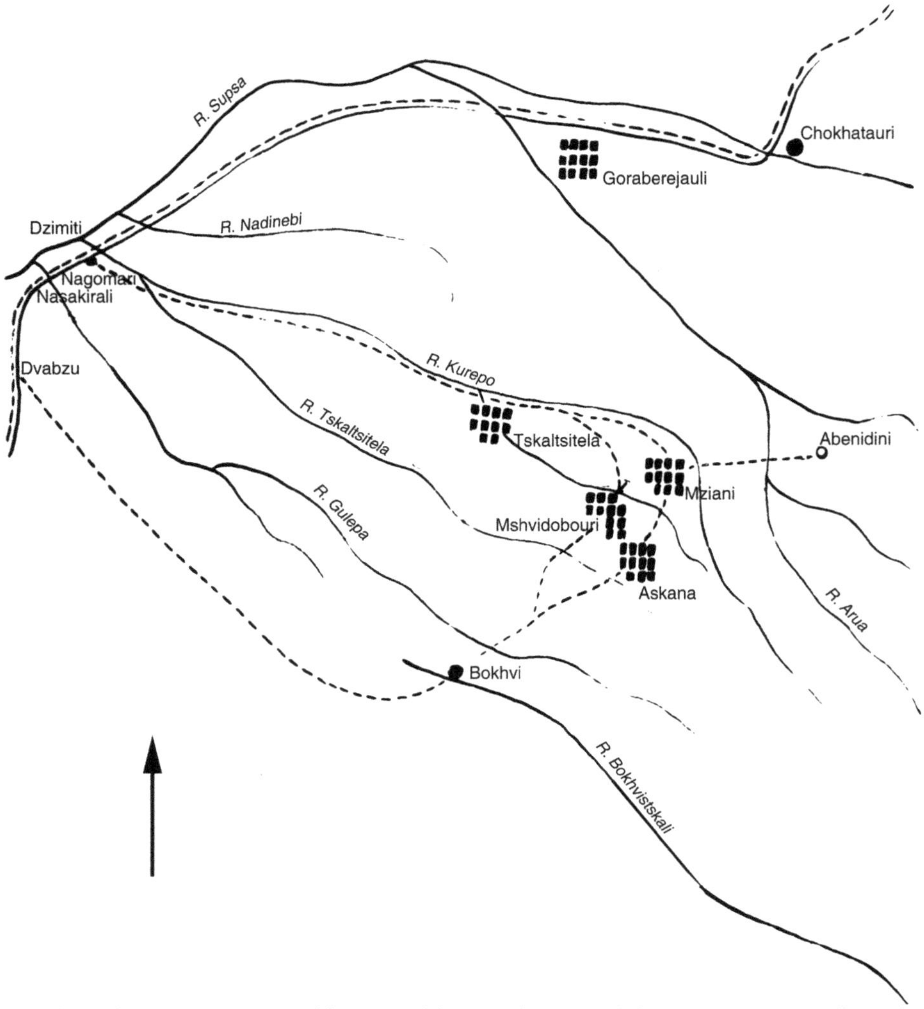

Fig. 20A. Location of separate groups of iron-smelting workshops of the Supsa-Gubazeuli production centre.

fixing the handle was usually a mining tool.[3] At the top of the dump was a store of fireclay kept under the dump for 'ripening' and for use when needed.

For determining the period when the furnace was used we had a stock of ceramic material similar to Colchian ceramics of the "Namcheduri III," "Leğva I," "Tsetskhlauri I" and "Tsetskhlauri II," "Tsikhisdziri IV" periods, etc. According to these data, the "Askana I" iron-smelting workshop may be assigned, approximately, to the tenth-ninth centuries B.C.[4]

2. IRON-SMELTING WORKSHOPS "ASKANA II"

The archaeological site "Askana II" consists of two iron-smelting workshops situated on the same plot, and two waste dumps on a slope facing south

The site is on a tea plantation belonging to the Mshvidobauri collective farm, at the source of the Tskaltsitela I stream, on the territory of the Daba-Etseri village, close to the homestead of farmer Minago Rizhamadze, on an elongated hillock bordering on the south by a little gully with a stream falling into the Tskaltsitela I.

a. Iron-Smelting Workshop "Askana II,1" (excavated in1974)

Of the iron-smelting workshop "Askana II,1"[5] there have survived remnants of the iron-smelting furnace, the base of the composite table (anvil) and the waste dump (Fig. 24).

The iron-smelting furnace "Askana II,1" has come down to us as a pit having the form of an upturned truncated pyramid. Of the stone facing only the lower part survives (70x50x60 cm) which has slipped down and rests on the hemispherical bottom of the furnace, evidently after it was no longer in use. The soil around the furnace has burned red by the heat to a depth of 15-20 cm. The

[3] J.I. Sunchugushev. *Ancient Mines and Monuments of Early metallurgy in the Khakass-Minusinsk Depression*. Moscow, 1975. pp. 68-71, Figs. 19-21.

[4] D.A. Khakhutaishvili. Production Centre of Colcho-Chalybean Mining and Metallurgical Area in the Supsa-Gubazeuli Gorge. In: *Monuments of South-Western Georgia*, X, Tbilisi, 1981, pp. 3-13, Figs. 1, 2. (in Georgian, summary in Russian).

[5] Preliminary publication: see ibid., pp. 13-21, Figs. 4-9.

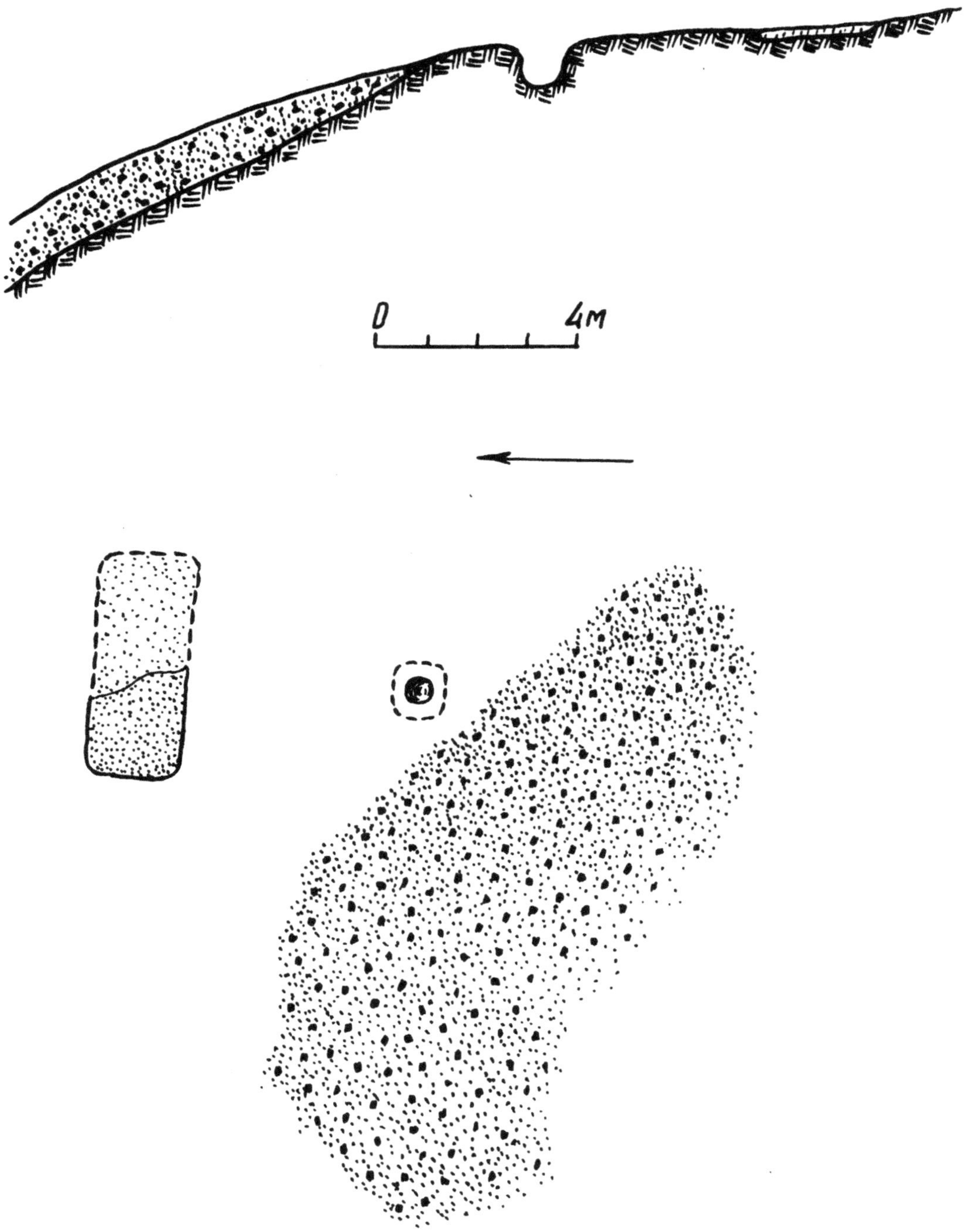

Fig. 21. Iron-smelting workshop "Askana I." General ground plan and section.

furnace was built on a flat part of the eastern part of the hill-top. The remnants of the furnace were covered by a layer of yellow diluvial clay. The soil here is the same as in all the subtropical foothill zone of the Eastern Black Sea littoral (a thick stratum of laterites beneath yellow clay). Inside the pit were lumps of clay plaster and slag, scorched stones from the structure, a considerable amount of charcoal, fragments of nozzles for bellows (Figs. 25, 4, 6), and ash. All this was, naturally, mixed with yellow clay.

The height (depth) of the pit is 130 cm, the width at the top 140 cm (without the facing), at the bottom 70 cm. The lower part of the furnace that has survived is lined with stone, has a quadrangular shape and a volume measuring 70×50×60 cm, ending in a hemispherical lower part.

Of the composite table (anvil) all that remains is a strip of red-burnt clay.

The waste dump occupies an area of almost 80 square metres on a south-facing slope of the hill. In the centre of the dump the cultural layer is about 100 cm thick. It contains a quantity of lumps of slag (about 8 cubic metres) with a thick greyish patina, fragments of nozzles for bellows and unidentifiable sherds of Colchian pottery vessels (Figs. 25, 1-3, 5), pieces of charcoal, ash and scorched stones from the structure of the furnace and the

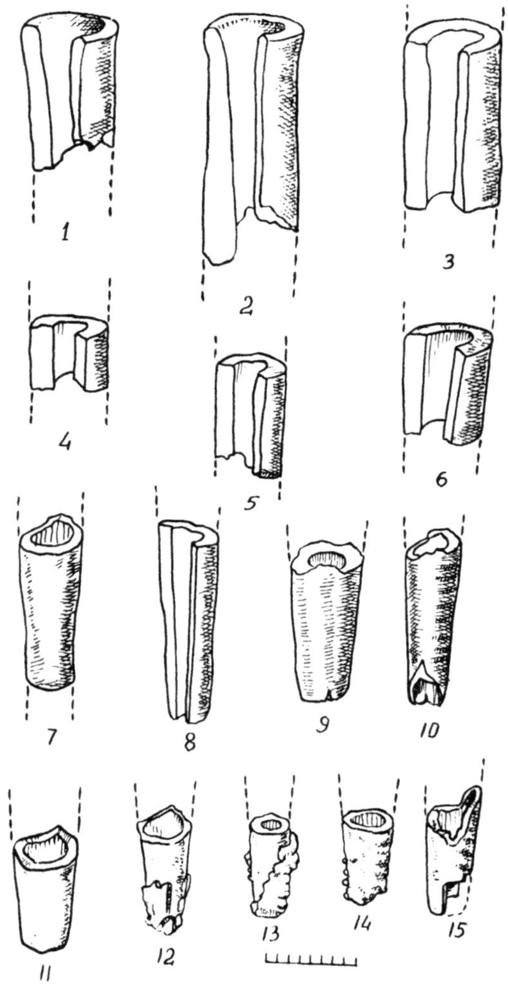

Fig. 22. "Askana I." Fragments of nozzles for bellows.

composite stone table. We should mention at once that we do not exclude the possibility but some material from the adjacent furnace somehow found its way into the waste dump.

There are certain data, according to which the time when the iron-smelting furnace "Askana II,1" was operating may be determined: a) the structure of the furnace, the aspect of the slag and the ceramic fragments (Fig. 25, 1-3, 5) attest that it belongs to the pre-Antique period; b) according to radiocarbon analysis, a sample of charcoal from the bottom of the iron-smelting furnace "Askana II,1" is dated to 1225 B.C. (3175±45),[6] which corresponds to 1420 B.C. (3370±25)[7] on the table compiled by Daymon *et al.* According to new archaeomagnetic data obtained by Z.A. Chelidze (declination = 64°, inclination = 14°, tension = 0.368 oersted), the furnace was used for the last time about the 14th-13th centuries B.C.

Thus, the iron-smelting furnace "Askana II, 1" functioned around the 15th-14th centuries B.C.[8]

b. Iron-Smelting Furnace "Askana II, 2" (excavated in 1974)

The iron-smelting furnace "Askana II, 2" is close to "Askana II, 1," to the west. The furnace itself, remnants of the composite table (anvil) and the waste dump have survived.

The iron-smelting furnace "Askana II, 2" is a small pit, an upturned truncated pyramid in shape, faced with stone. The ground, at a depth of 15-20 cm around the pit is burnt red by the heat.

The furnace was built in the western part of a hilltop, 15 metres from "Askana II, 2." The underground part of the furnace was found in good condition; its dimensions are: height (depth) 70 cm; width at the top 65 cm; in the middle 30 cm; at the bottom 15-20 cm. The inner space is filled with various-shaped stones, evidently

6 Burchuladze and Togonidze. *Radiocarbon Dating*, IV, Tbilisi, sample TB-234.
7 Arslanov. *op. cit.*, pp. 11-61.

8 Khakhutaishvili. *On the Chronology....*, p. 140; idem, *Production Centre....* p. 17.

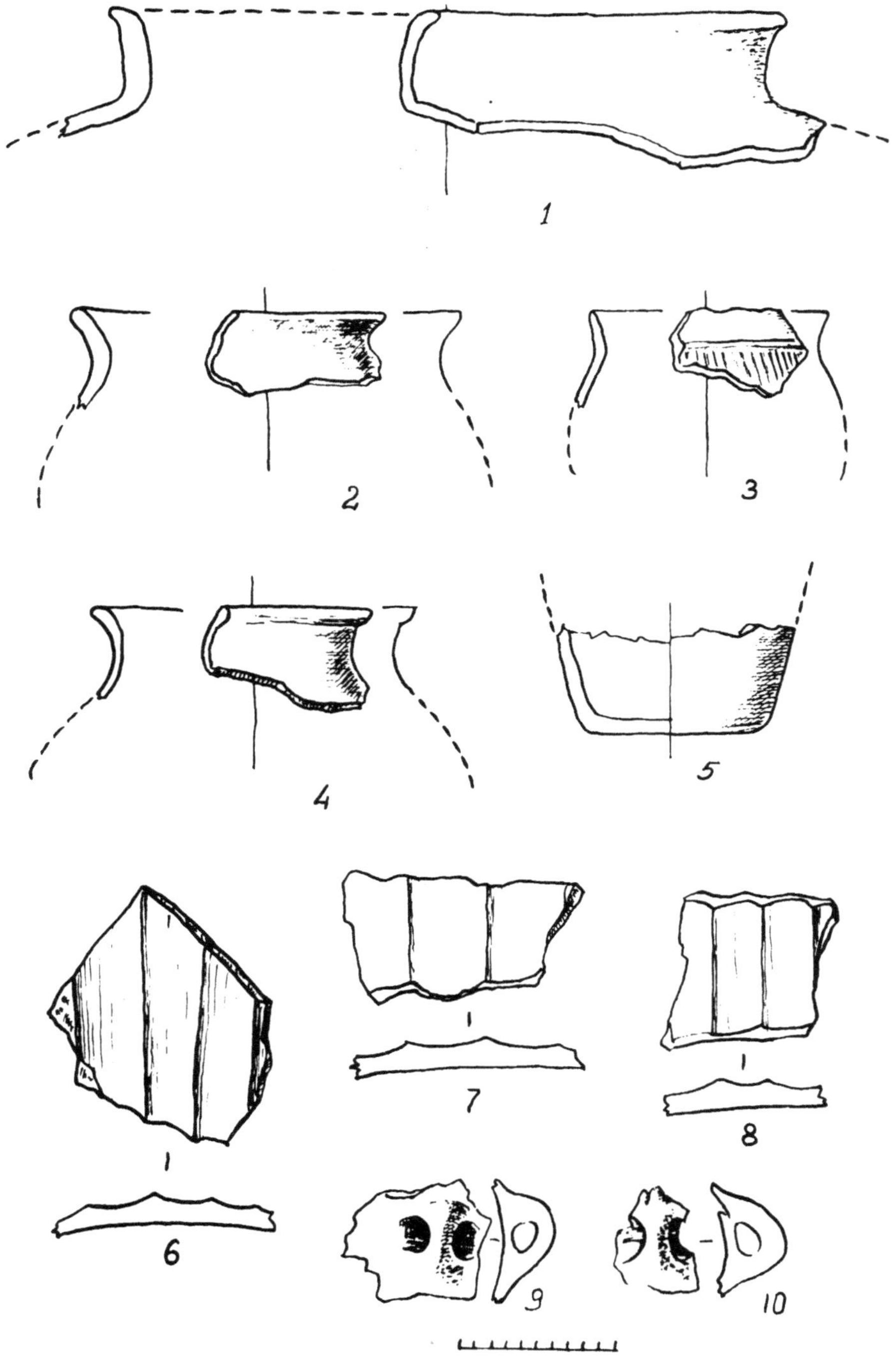

Fig. 23. "Askana I." 1-5 – fragments of clay pots; 6-8 – fragments of thick-walled burnished clay vessels ornamented with broad fluting; 9-10 – lugs from vessels.

from the structure of the part of the furnace above ground; lumps of slag and fragments of plaster, ash-grey in colour; clay burnt red, potsherds, including the flat bottom and a rim of a Colchian ceramic vessel, as well as fragments of nozzles for bellows (Fig. 26). At the bottom of the furnace was a layer of charcoal.

Of the composite stone table there remains an area of 200×100 cm burnt red and some scattered scorched stones.

The waste dump covers about 100 square metres of ground on a slope of the hill. The cultural layer here is up to 120 cm. In it were found numerous lumps of iron slag covered with a thick grey patina (about 10 m^3), pieces of clay plaster of the furnace, ash-grey in colour (about 4 m^3), fragments of nozzles for bellows and sherds of Colchian pottery (Fig. 26) made on a potter's wheel from clay with an admixture of magnetite. The lower part of the dump stretches down to the bed of the gully and is covered with a thick layer of silt in which tea shrubs have been planted.

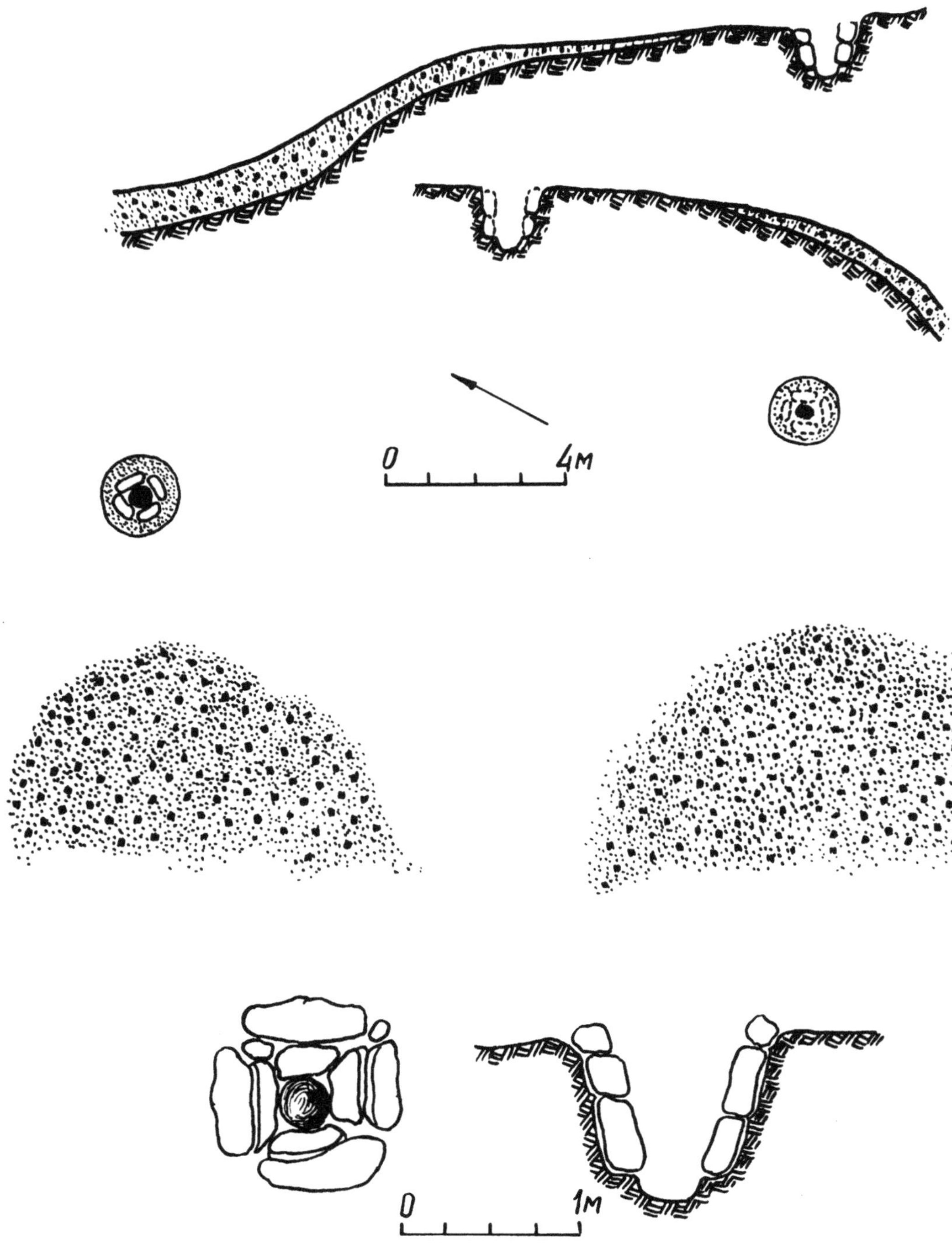

Fig. 24. Iron-smelting workshops "Askana II." General ground plan and sections.

The potsherds and clay are not informative enough in the matter of dating the site. They only point to pre-Antique times. According to radiocarbon data obtained from a study of samples from the bottom of the iron-smelting furnace, it functioned around 1130 B.C. (3080±45), corresponding to 1312 B.C. (3262±125)[9] by the table compiled by Daymon *et al.*

According to the latest archaeomagnetic data obtained by Z.A. Chelidze, the iron-smelting furnace "Askana II, 2" was used for the last time in the 14-13th centuries B.C.

3. IRON-SMELTING FURNACES "ASKANA III"

The site with remains of ancient iron production, "Askana III," was identified in 1975. This site had a place of its own among those previously found. The fact is that of three iron-smelting workshops found here, only one belongs to the pre-Antique period; as for the two others, all the data attest that they were functioning in the Early Antique period.

[9] Burchuladze and Togonidze. *op. cit.*, p. 14. Sample TB-235. Ch.A. Arslanov. *op. cit.*, pp. 11-61. Cf. Khakhutaishvili. *On the Chronology...*, p. 140; idem.. *Production Centre...*, p. 21-30.

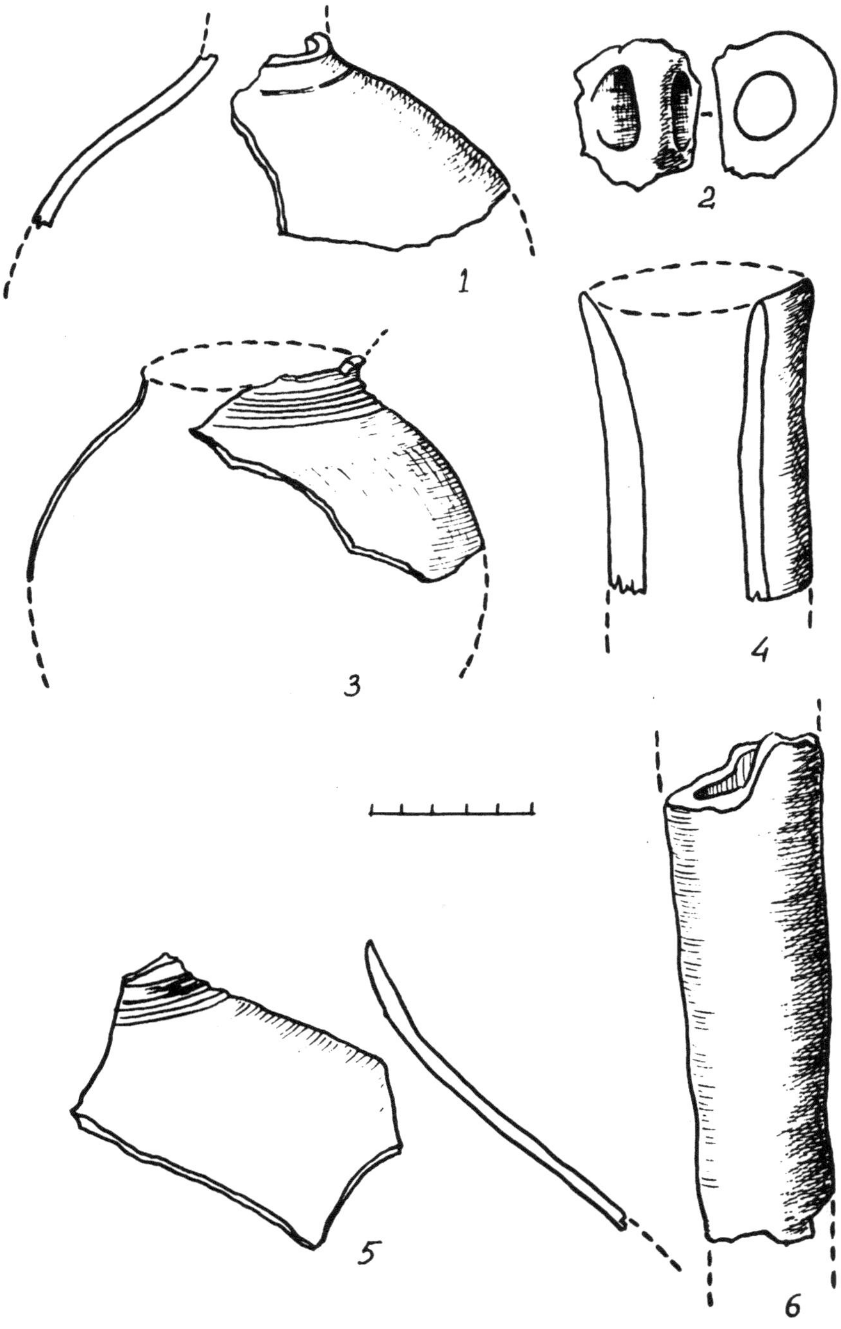

Fig. 25. "Askana II, 1." 1,3,5 – fragments of clay vessels; 2 – lug from a vessel; 4-6 – Fragments of clay nozzles for bellows.

The south-facing site is situated on a broad triangular slope and belongs to a chain of hills, the watershed between the Tskaltsitela I and Tskaltsitela II, north of the tea-processing factory in the village of Dabali Etseri, 1 km from the Askana - Dabali Etseri - Nagomari road. The slope on which the remnants of workshops are situated is limited on the west and south-east by deep gullies which converge at the foot of the slope making a promontory-like relief sharply falling in the direction of the gullies. The upper part of the slope was occupied by a tea plantation, the lower part was overgrown with shrubs and fern. Three waste dumps were identified on the surface of the slope, different in size and outward appearance. One dump, more ancient in aspect, was on the higher part of the slope, at the edge of the plantation; the two others on the lower part, on the slopes of the two gullies.[10]

[10] For preliminary publication of the site see: Khakhutaishvili. *Production Centre...*, pp. 21-30, Figs. 10-15.

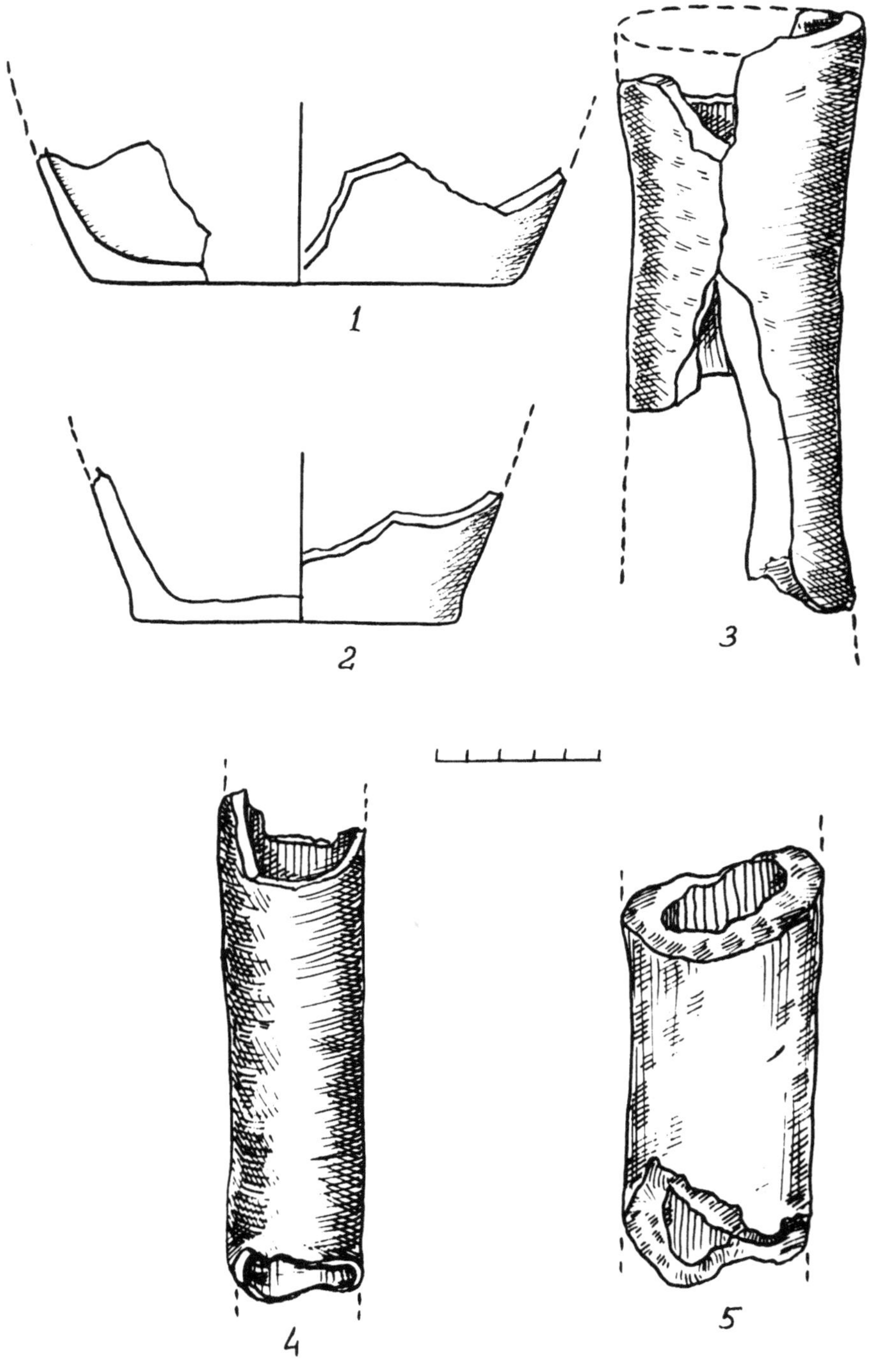

Fig. 26. "Askana II, 2." Bottoms of vessels; 3-5 – fragments of flute-like clay nozzles for bellows.

a. Iron-Smelting furnace "Askana III, 1" (excavated in 1975)

Nothing has survived of the "Askana III, 1" iron-smelting workshop but the waste dump. The furnace itself and the composite table (anvil) for working the bloom were destroyed when the Mshvidobauri village tea plantation was being laid out. The waste dump of "Askana III, 1" iron-smelting workshop covers a rather large space (30x20 metres) where the depth of the cultural level attains, in places, 40 cm. In the dump, a large amount of iron slag was found (about 8 cubic metres), the lumps being covered with a thick greyish patina; also sherds of pottery, including fragments of pots and other vessels with various zoomorphic handles (Fig. 27), conical nozzles with a funnel-shaped opening (Fig. 28). One of the nozzles bears a representation of a snake (Fig. 28, no. 16), on the top part of another there is a deep groove for the fixing cord with which the nozzle is attached to the orifice of the leather bellows (Fig. 29 no. 2). It is the first time that fragments of this kind of nozzle have been found, and they help us to gain an idea of the structure of bellows for bloomery smelting. As regards the representation of a snake, we may note that according to ethnographic and folklore data collected in the locality and recorded in literature, one variety of snake, was called "the family snake" and was regarded as the patron protector of the family and its wellbeing; a deliberate killing of a snake was supposed to bring misfortune. In the material culture of the whole territory inhabited by Kartvelian (Georgian) tribes, the representation of snakes occurs from early times.[11] In the waste dump, another find was a portion of a crucible (Fig. 29, 1) with traces of smelted metal. The ends of some nozzles of bellows, filled with slag, deserve attention (Figs. 28, 14, 15, 17-19).

In order to date this workshop, a most expressive group of fragments of Colchian earthenware vessels may be used (Fig. 27). Such pottery often occurs in the material from Colchian sites.

b. Iron-Smelting Workshop "Askana III, 2" (excavated in 1975)

The remnants of this workshop have survived below the waste dump of the workshop "Askana III, 1," 50 metres away from it, at the edge of the nameless western gully. It was the appearance of the dump that drew our attention, more exactly, the aspect of the slag the lumps of which were not covered with thick grey patina, as is usually the case, but with a thin coating of rust. Moreover, some fragments of the clay plaster of the furnace were not ash-grey in colour, but brick-red. These features observed during a routine examination of the dump dictated especial care when excavating. Unfortunately, major landslides and great changes of the relief had almost completely destroyed the "Askana III, 2" workshop: of the iron-smelting furnace there remained a half-sunk badly scorched oblong pit, which makes it impossible to gain an idea of the structure of the furnace; the composite table (anvil) has been swept away into the gully, as well as the waste dump. Excavation of what remained of the dump yielded various lumps of slag and pieces of plaster, fragments of nozzles and shapeless sherds of ancient Colchian pottery fired mostly red or brown. Such pottery was widespread in the Eastern Black Sea littoral in the Early Antique period.

c. Iron-Smelting Workshop "Askana III, 3" (excavated in 1975)

"The "Askana III, 3" was placed not far from the "Askana III, 2" workshop, to the east of it, near a dry nameless gully which defined the slope on the west. The workshop consisted of the iron-smelting furnace, the composite stone table (anvil) and the waste dump.

The iron-smelting furnace "Askana III, 3" (Fig. 30) has come down to us as an oblong pit dug out of a sharply sloping surface and faced on three sides with roughly hewn stones and cobbles. The front wall of the furnace was a temporary structure: it rested on the stone floor of the furnace and, most probably, was removed after each smelting or an orifice was made in it to take out the slag. At the back wall of the furnace there is a hemispherical cavity, the technical purpose of which is not quite clear.

The back wall of the furnace is in good condition, of the two sides there remains a part, along the line of the slope. The furnace is 140 cm long, 30-40 cm wide, 100 cm high (at the back wall); judging by the parameters of other parts of the furnace, its height could not have exceeded 150-160 cm.

The ground (yellow clays) around the furnace was reddish from the heat to a thickness of 30-35 cm; the surface of some stones of the facing was melted.

The inner space was filled in with scorched stones from the structure of the furnace, and with fired clay. On the stone floor of the furnace we cleared several lumps of highly plastic, well-refined bluish clay.

Thus, "Askana III, 3" smelting furnace is very different in structure from pre-antique Colchian furnaces.

Of the composite table (anvil) there remain a few stones and a small plot (150×100cm) burnt red, most probablythe base of the anvil.

The waste dump was on the steep slope of the gully, consequently it contained a comparatively small number of lumps of slag, pieces of clay plaster and nozzles for bellows. Here, too, some slag lumps have a coating of reddish rust, and the fragments of clay plaster are brick-red in colour. The potsherds found in the dump were burnt red and brown. These data prove that the iron-smelting workshops, "Askana III, 2" and "Askana III, 3" to belong to the same group. The contemporary nature of these sites is also attested by fragments of pottery, in particular potsherds fired red and brown, characteristic of the Early Antique period.

Thus, the excavations of the archaeological site "Askana III" proved that this Colchian centre of ancient iron metallurgy also functioned in the Antique period. Besides, the excavation of this site made it possible to determine one of the principal stages of the development of iron-smelting furnaces in the Eastern Black Sea littoral.

4. IRON-SMELTING WORKSHOP "ASKANA IV"

The iron-smelting workshop "Askana IV"[12] was identified and excavated in 1975. The remnants of this

[11] Incidentally, we might recall in this connection a legend told by Apollonius of Rhodes that the Golden Fleece was "guarded by a Serpent that wound itself round the Fleece, unyielding to death or to sleep; It was born of Gaea in the Caucasus... Farther on was the field of Ares, and the god's sacred grove where the Serpent guarded the Fleece watchfully". Apollonius of Rhodes, *Argonautica*. Translated into Georgian by G.Th.Tsereteli. Tbilisi, 1964, pp. 132-133, 137.

[12] For a preliminary publication of the site, see: Khakhutaishvili. *A Production Centre ...*, pp. 30-35, Figs. 16-20.

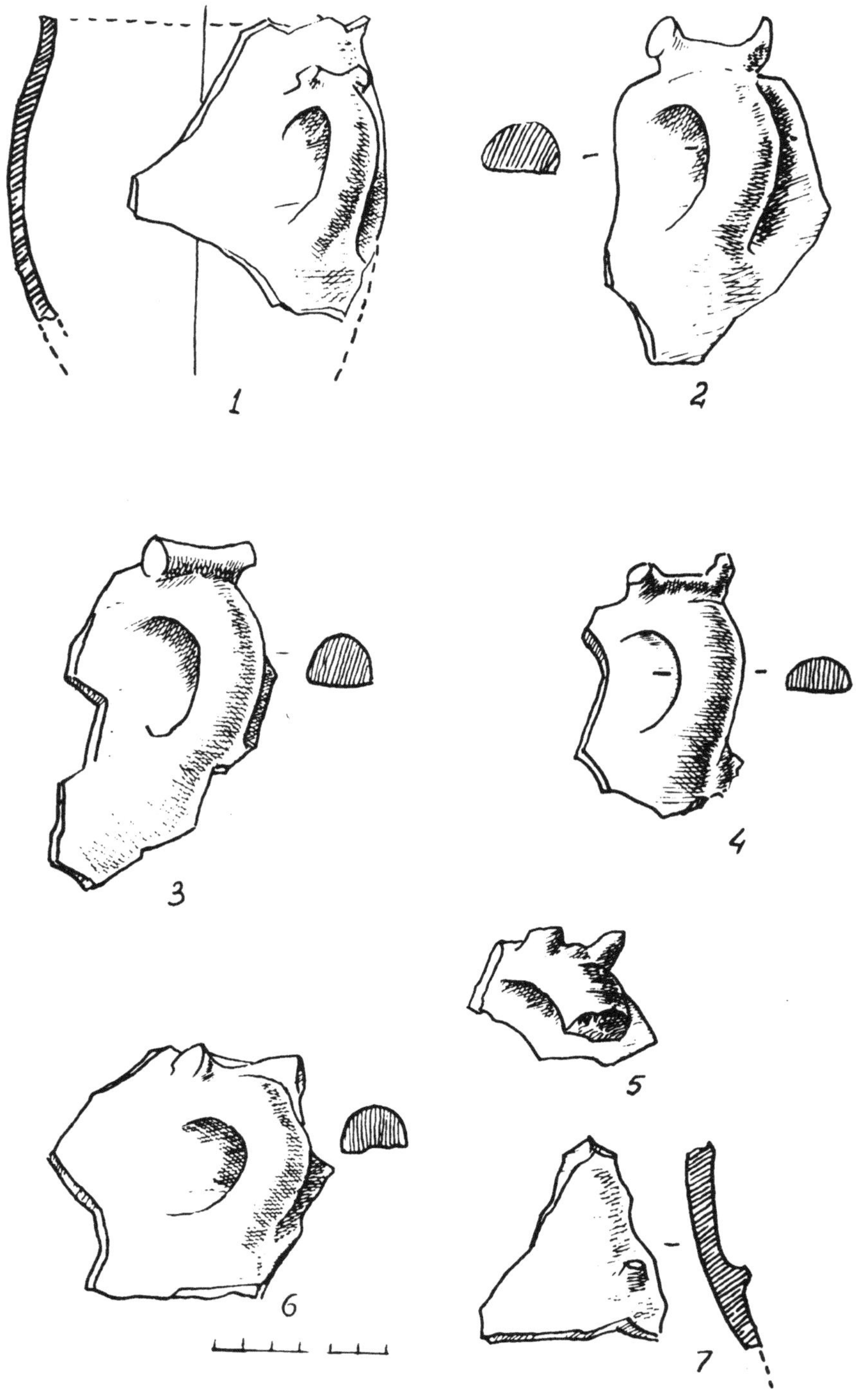

Fig. 27. "Askana III, 1." 1-6 – fragments of vessels with zoomorphic handles; 7 – fragment of a vessel.

workshop which survived in the village of Askana, Moskaneti area, to the west of Giorgi Ninidze's holding, are to be found on a hill stretching from north to south with a high-voltage electricity pylon on it. Remnants of the workshop were covered by a layer of clay 35-40 cm thick.

The workshop was located by its waste dump on the west slope of the hill.The workshop consisted of an iron-smelting furnace, a composite table (anvil) for forging the bloom and a waste dump (Fig. 31).

The iron-smelting furnace "Askana IV" was a small pit, shaped like an upturned truncated pyramid with a hemispherical bottom and faced with roughly hewn stones and cobbles (Fig. 31). From the effect of high temperature, the stones of the facing had become lilac-coloured in places.

The dimensions of the furnace; height (depth) 85 cm; width at the top 40 cm, below, at the edge of the hemispherical depression - 20 cm.

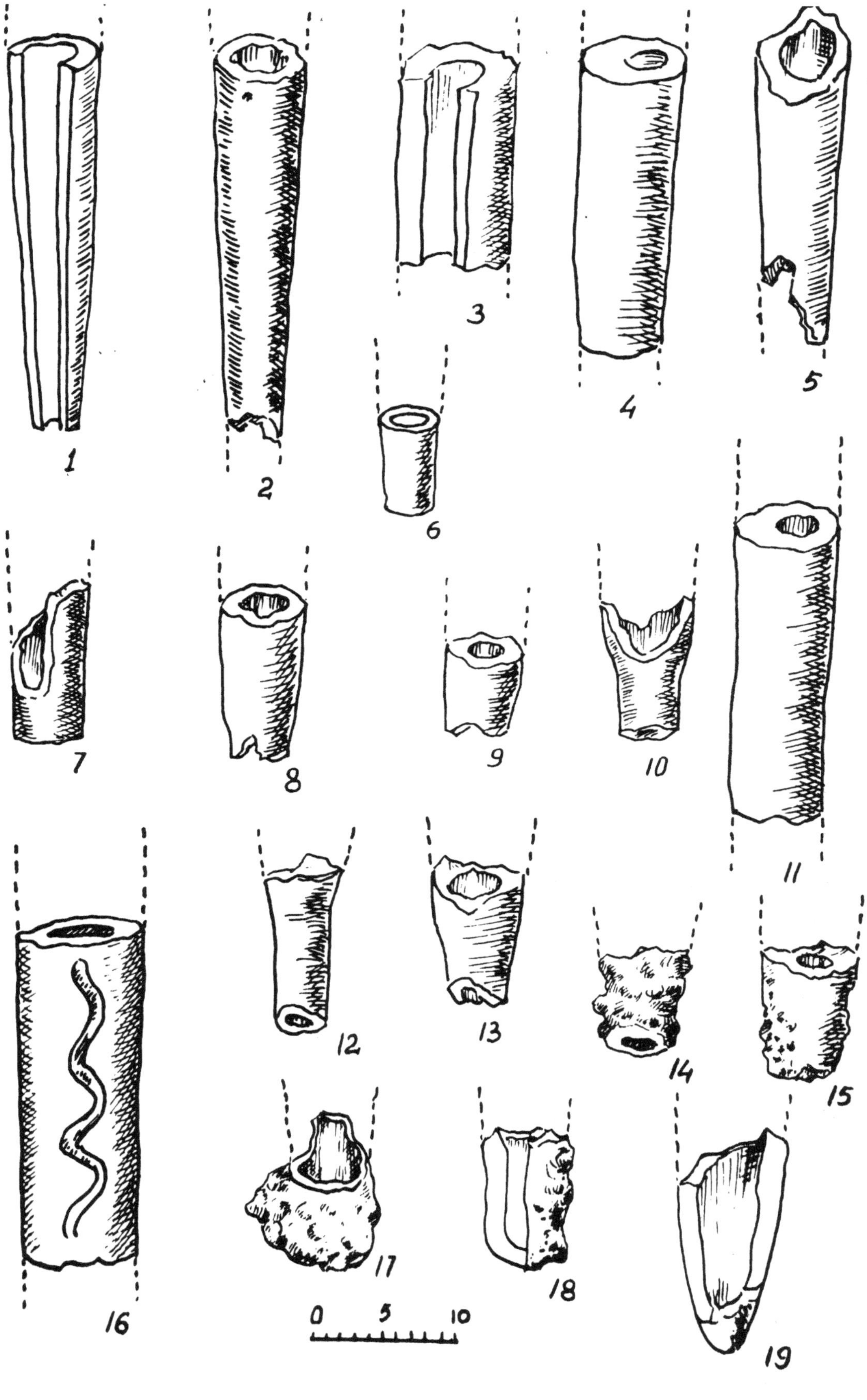

Fig. 28. "Askana III, 1." 1-19 – fragments of clay nozzles for bellows;
16 – fragment of a nozzle with a snake design fashioned in relief.

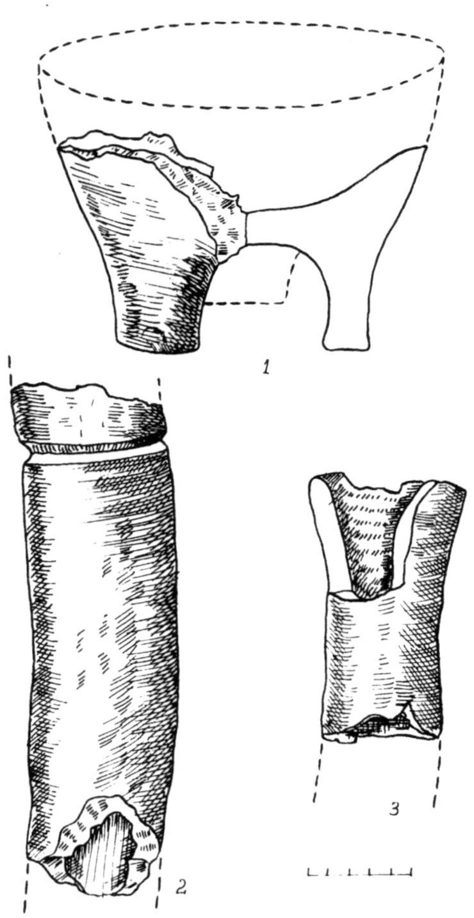

Fig. 29. "Askana III, 1." 1 – fragment of a clay crucible; 2 – fragment of slightly curved nozzle with a groove; 3 – fragment of a nozzle for bellows.

Inside, the furnace was filled with scorched stones from the structure of the furnace, lumps of iron slag and fragments of nozzles for bellows (Fig. 32), crumbled charcoal, ash, burnt clay, etc. In the material from the furnace two fragments of nozzles are noteworthy (Fig. 32, 4-5). One of the fragments belongs to a flute-like nozzle (Fig. 32, 4). A fragment of this kind was found in material from the "Tsetskhlauri I"[13] smelting furnace (V. above).

The other fragment is bent and was intended to be a transition of the line of air supply from the surface to the inside of the furnace. As has already been mentioned, we have never recorded a case of air being supplied from the bottom or from the walls of the underground part of iron-smelting furnaces.

The soil around the furnace, consisting of yellow clay was burnt red by the heat to a depth of 15-20 cm.

The composite table (anvil) round in shape, with a diameter of 100 cm, consisted of cobbles (Fig. 3l). The area around the table was reddish in colour because of the high temperatures and by the oxidization of iron particles sprung from the spongy bloom under the impact of the blows of the beetle.

The waste dump was on the west slope of the hill. Unfortunately, all that remained of it was a few badly damaged portions of it and several small heaps of slag, collected by the peasants when clearing the ground of hard lumps of iron slag, stones, etc. littering it. Excavations of the remains of the cultural layer yielded some lumps of iron slag covered with a thick grey patina, ash-grey fragments of clay plaster, as well as fragments of nozzles. On the slope, some objects were found: part of a mortar fashioned out of a flat cobblestone, with two hollows for breaking up the flux (Fig. 32, 1) and two pestles made out of pieces of iron slag and of a large rounded pebble (Fig. 32, 2, 3).

There is no material to suggest the date of the workshop. According to the structure of the furnace and the aspect of the slags (a thick patina, grey in colour and characteristic of slags of the pre-Antique period), this shop must have functioned in the pre-Antique epoch. According to archaeomagnetic data obtained by Z.A. Chelidze, this workshop belongs to the ninth or 14th-13th centuries B.C. (declination= 8°, inclination = 74° tension = 0.353 oersted).[14]

IRON-SMELTING WORKSHOPS OF THE "MZIANI" GROUP

The group of iron-smelting workshops "Mziani" was identified in 1975 on the territories of a village of the same name (formerly called Pichkhis Djvari), in the Ozurgeti district of the Georgian SSR. It belongs to the central group of sites of the Supsa-Gabezeuli production area and comprises over 30 units.

The village of Mziani is located north of Askana village, separated from the latter by a small nameless gully limiting Mziani on the south. The gully empties into the Kurepa brook, a right-hand tributary of the river Gubazeuli which, in its turn, falls into the Supsa river near the village of Nagomari. It is basically the Supsa and its principal tributaries that supply the Ureki beaches with magnetitic sands.

The territory of the village of Mziani, with all its arable land is limited to the south by the territory of Askana village, to the north by that of the villages of Buknari and Nagomari, to the west by Magali Etseri and Mshvidobauri villages, and to the east by the lands of the villages Basileti and Kvenobani (along the Kurepa river). According to our observations, the iron-smelting workshops identified on the territories of the villages: Askana, Mziani, Mshvidobauri, Nagomari, Dabali, and Etseri formed the nucleus of the Supsa-Gabezeuli iron-producing area.

[13] D.A. Khakhutaishvili. *Materials on the History ...*, pl. VI, 2.

[14] Z.A. Chelidze. *Results of Archaeomagnetic Investigation of Some Archaeological Sites*, p. 148.

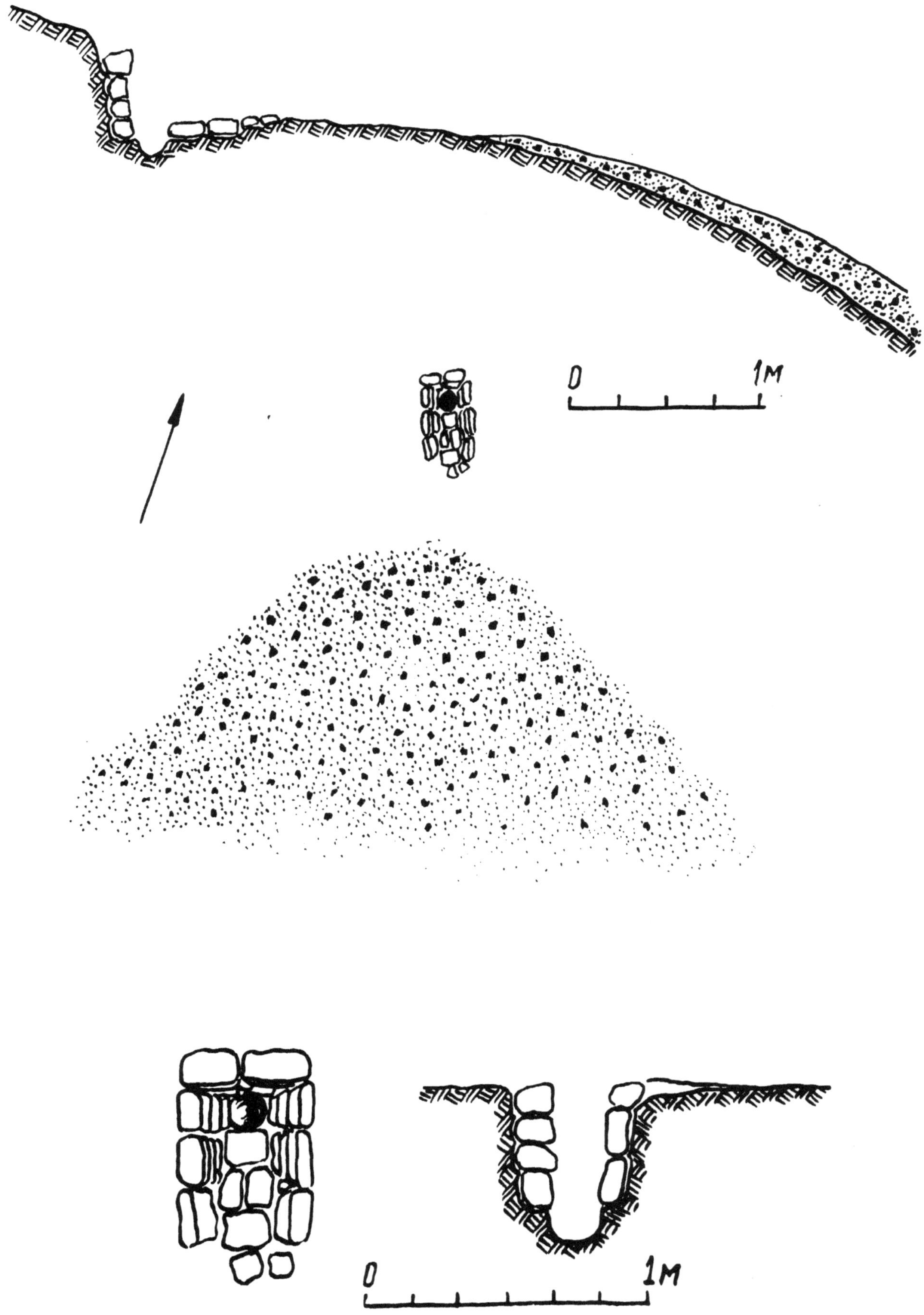

Fig. 30. Iron-smelting workshop "Askana III, 3." Ground plan and sections.

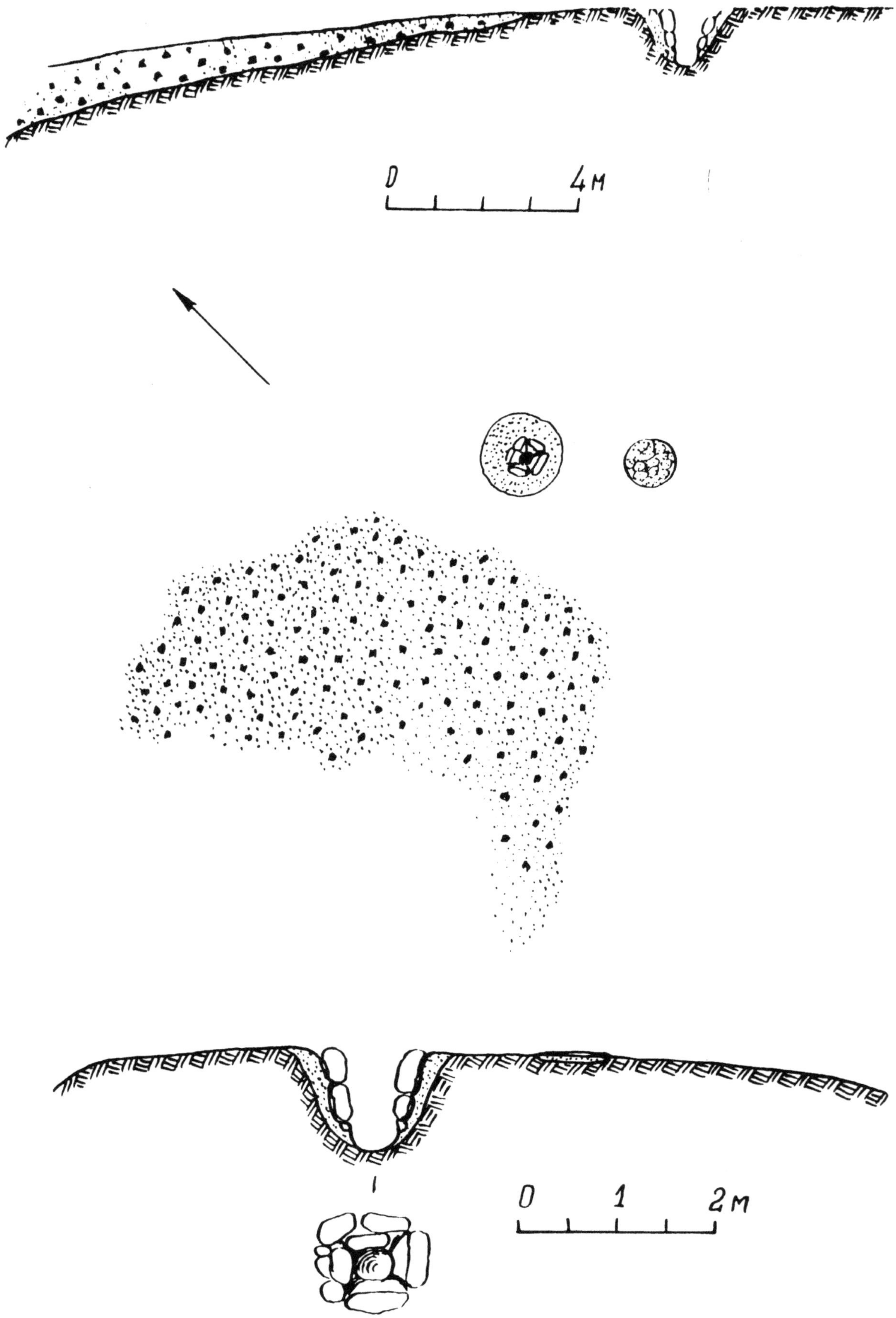

Fig. 31. Iron-smelting workshop "Askana IV, 3." Ground plan and sections.

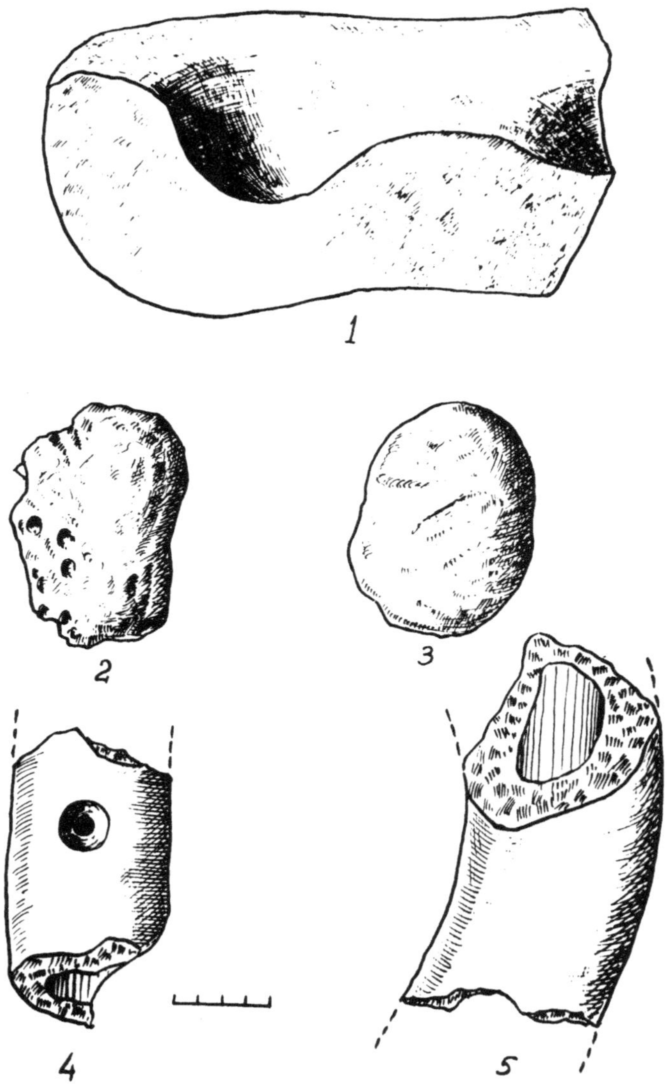

Fig. 32. "Askana IV." 1 – fragment of a mortar for crushing flux; 2,3 – pestles; 4 – fragment of a flute-like nozzle for bellows; 5 – fragment of a curved nozzle for bellows.

The geological and soil conditions here are the same as in all the other centres of iron manufacture in central and southern Colchis: the same subsoil of tufogenic Tertiary laterites overlaid by yellow clay of diluvial origin, plentiful fuel and rugged terrain with a thick network of brooks and streams. Most of the sites connected with the manufacture of iron are concentrated on the left bank of the Kurepa river.

Identification was greatly aided, as in other cases, by such indexes as the waste dumps scattered over the hills, on the banks of streams in the villages, or on the lands of villages.

5. IRON-SMELTING WORKSHOP "MZIANI I" (excavated in 1983)

The iron-smelting workshop "Mziani I" (Fig. 33) was identified in 1975 by the expedition of N.A. Berdzenishvili, Batumi Research Institute, Academy of Sciences of the Georgian SSR.[15]

The remains of the workshop were found on a south-facing slope, to the south-west of R.S. Djincharadze's holding, on the left side of a nameless gully on the boundary of the Askana and Mziani village territories.

Judging by the volume of the waste dump, the workshop was in action for a long time, but, unfortunately, we failed to locate the furnace (or furnaces). Before the October revolution a peasant's house stood there, and it is highly likely that the furnace was destroyed because reconstruction of the homestead required digging the plot. On the other hand, a road was laid across the plantation and across the plot where the workshop had been, which

[15] The personnel of the expedition: D.A. Khakhutaishvili (head), S.I. Gogitidze (senior researcher), A.M. Djavelidze (artist), R.K. Atoyan (topographer), E.G. Kvirikadze (photographer), M.A. Zarkua (driver).

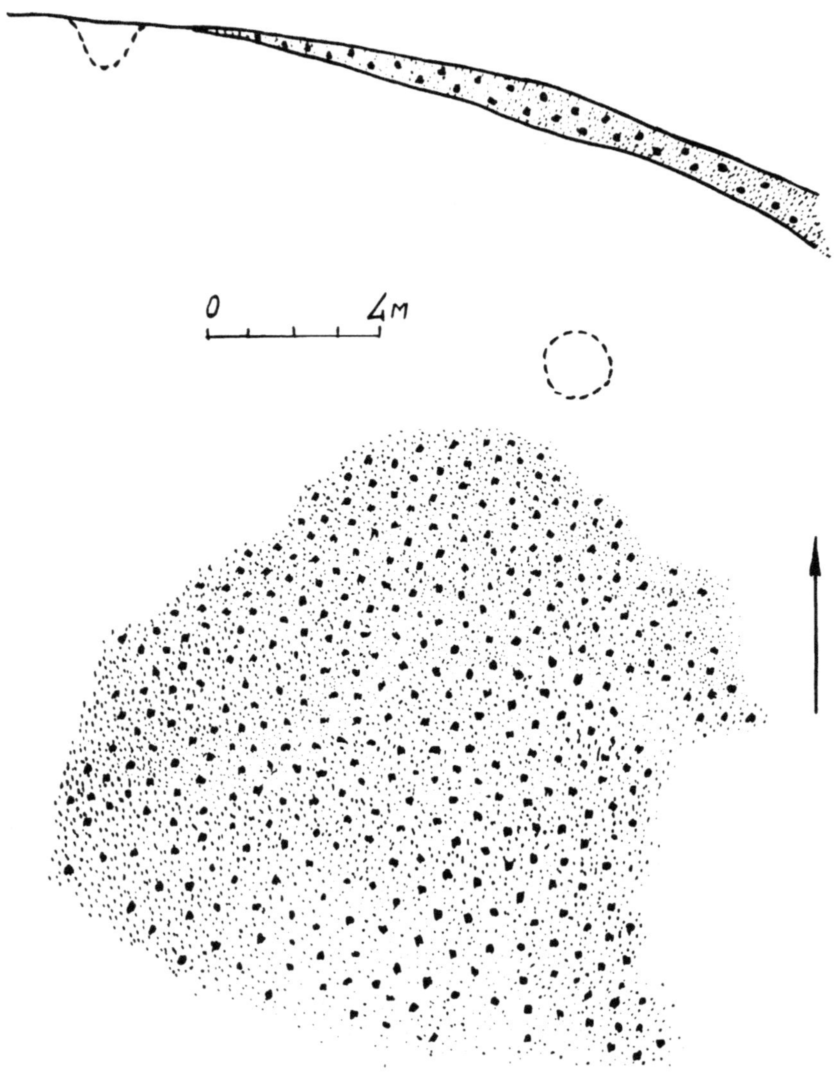

Fig. 33. Remains of iron-smelting workshop "Msiani I."

may also have damaged the other parts of the workshop, including the composite table (anvil) for forging the bloom. Thus, the only thing that was left of the workshop was the waste dump, which covered about 300 m^2; but even that had been impaired: according to local residents, the owner of the holding had been "battling" with that dump all his life, to give his place a more "decent" look and to gain more profit from it. Evidence of his activities is there: several heaps of iron slag, fragments of plaster and other iron production waste.

Excavation of the waste dump yielded a considerable amount of slag (about 10 cubic metres) covered with a thick greyish patina, of clay plaster from the furnace (about 3 m^3), fragments of clay cone-shaped nozzles for bellows (Fig. 34), mortars for pounding flux (Fig. 36), potsherds of vessels of different times (Fig. 35): among them there is a fragment of a Colchian cup (Fig. 35, 9) and of a thick-walled vessel on which a Greek υ (upsilon) is scratched. In all probability, part of a monogram (Fig. 35, 6); the zoomorphic handle of an earthenware pot (Fig. 35, 10), the flat bottom of a clay vessel and the fragment of a vessel decorated with narrow fluting (Fig. 35, 7, 8) may be assigned to the eighth-seventh centuries B.C. Sherds of thick-walled ceramic vessels decorated externally with broad fluting done by the "cutting" technique are noteworthy. As has been mentioned above, such vessels occur in the strata of the second millennium B.C. down to the eighth century B.C.[16] throughout maritime Colchis.

Thus, going by archaeological materials, the "Mziani" workshop functioned for a considerate time, over several centuries and most probably had more than one iron-smelting furnace. In future, when we have iron slag, the suggested chronology will be verified.

[16] D.A. Khakhutaishvili, *On the Chronology...*, pp. 136-139.

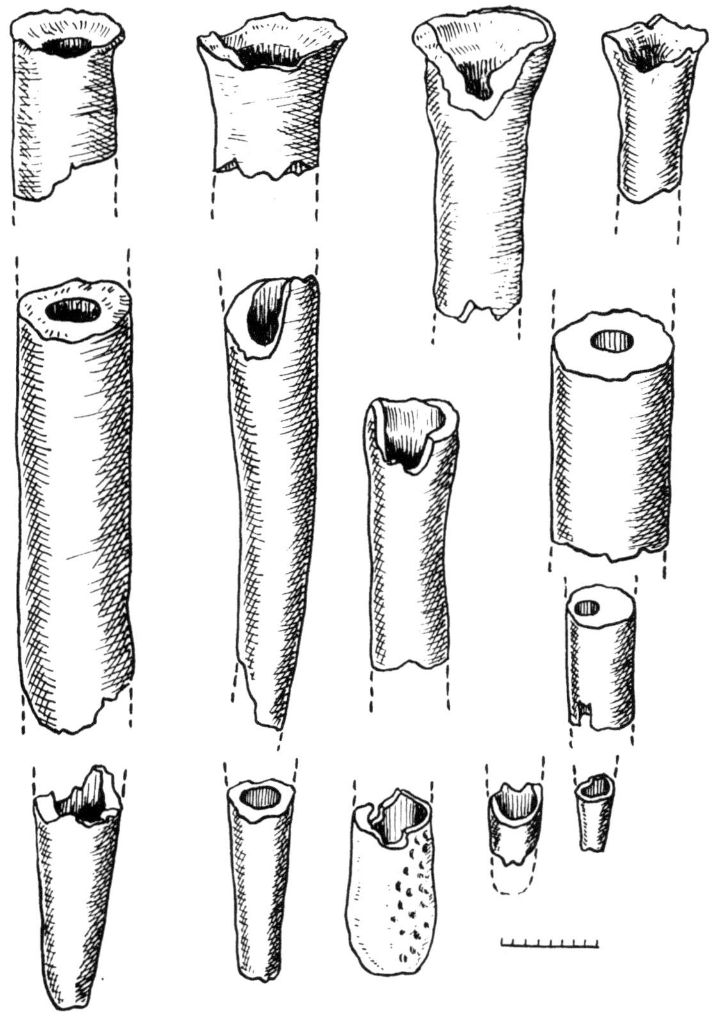

Fig. 34. "Msiani I." Fragments of clay nozzles for bellows.

6. IRON-SMELTING WORKSHOP "MZIANI II" (excavated in 1983)

The iron-smelting workshop "Mziani II" was located, (judging by what remains) on a promontory-like hill at the confluence of two nameless gullies belonging to the hydro system of the Kurepa river. The workshop was identified by its waste dump on the southern slope of the hill at the top of which is a small plot planted with tea. This plot is adjacent to a slope with a southerly exposure, partly occupied by a tea plantation. The remains of the workshop are near a peasant homestead owned until 1975 by the farmer Nikola Gogoladze, and since 1976 by the farmer Guram Mikadze. The tea plantation belongs to the Askana village collective farm which unites the villages of Askana, Mziani, Dabali Etseri, Mağali Etseri and others.

The remnants of the workshop comprise the iron-smelting furnace, the composite table (anvil) and the waste dump (Fig. 37).

The iron-smelting furnace "Mziani II" (Fig. 37) consists of a pit shaped as an upturned truncated pyramid and covered by a sterile layer of diluvial yellow clay 25-30 cm thick. The pit was faced with stones, of which only the lower course remains; above it, at a height of 50 cm, there are grooves of the facing stones. The interior of the furnace was lined down to the very bottom for the first time in our experience with a single flat stone. There were signs that the whole interior was plastered with a thick layer of fire-clay.

The height of the surviving part of the furnace is 125 cm, the dimensions are, at the top, 125×115 cm, at the bottom 48×43 cm.

At a depth of 15-20 cm the ground around the furnace was burnt red by the heat. At the western edge of the furnace there was a pit 70×80×15 cm in size, that eas full of crushed charcoal and scorched clay (Fig. 37). During clearing the zoomorphic handle of a large clay vessel was found. This pit had probably been used for setting up the bellows; more exactly, it was the place where the leather body of the bellows was joined to clay nozzles.

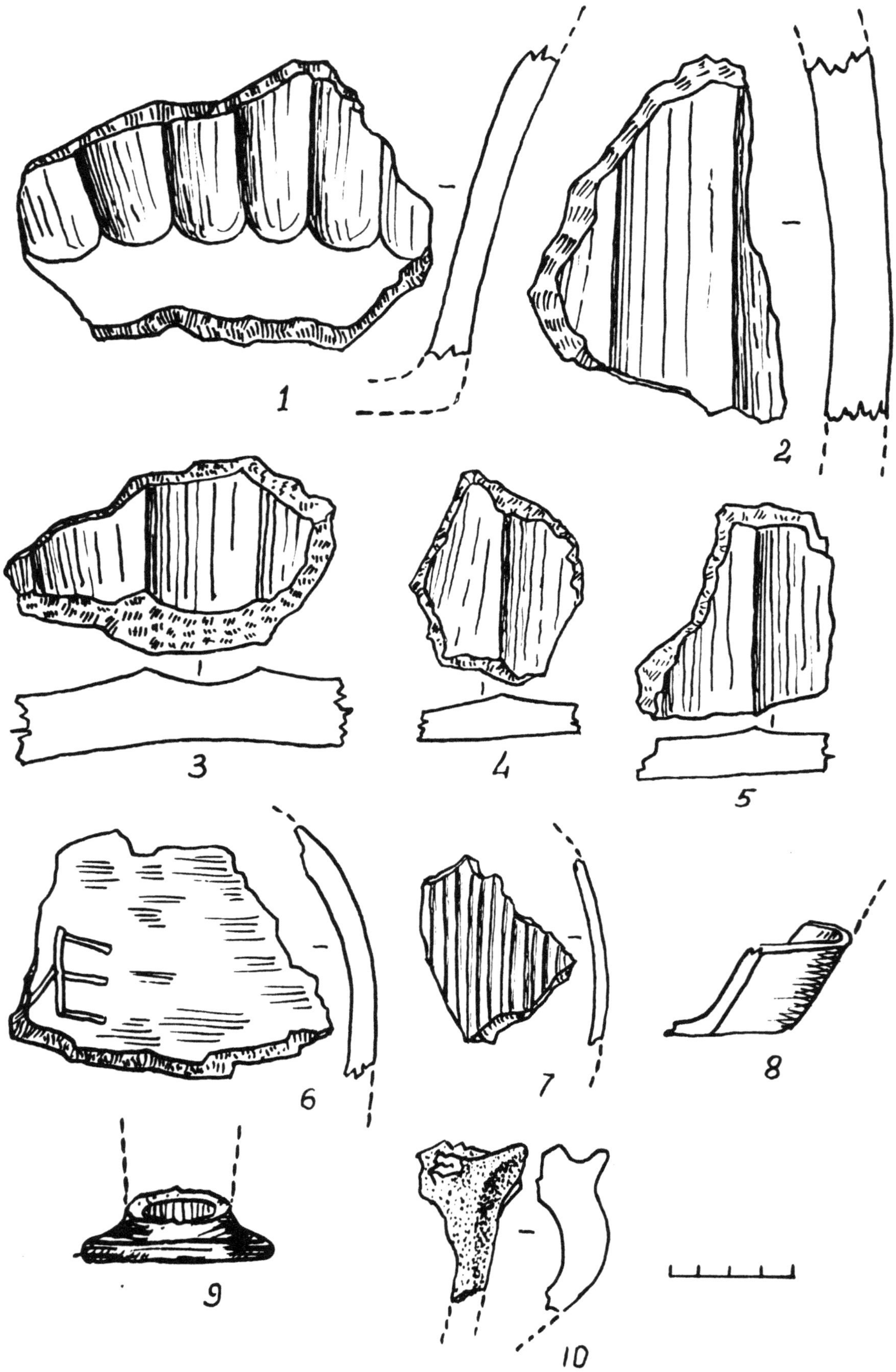

Fig. 35. "Msiani I." 1-5 – fragments of thick-walled burnished clay vessels ornamented with broad fluting; 6 – fragment of a vessel wall with graffiti; 7 – fragment of a vessel with narrow fluting; 10 – fragment of a zoomorphic handle of a vessel.

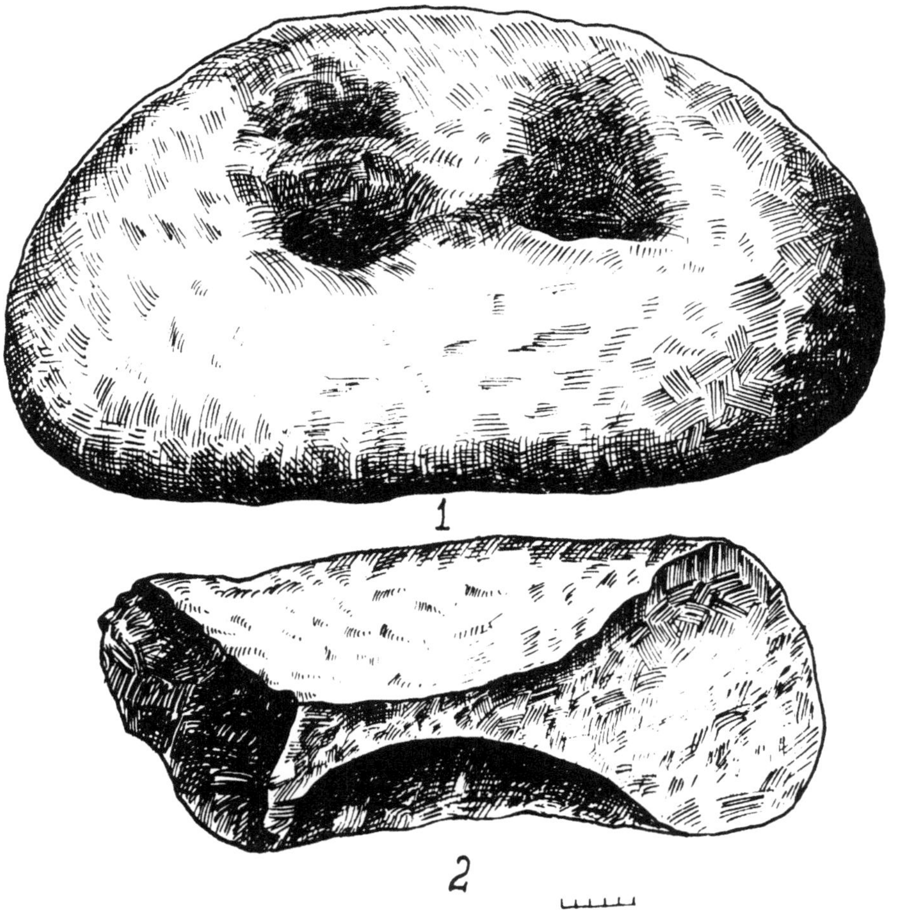

Fig. 36. "Msiani I." 1 – mortar for crushing flux; 2 – mortar fragment.

The inner space of the furnace was filled with scorched stones from the structure of the furnace facing, clay, crushed charcoal, lumps of slag, fragments of nozzles and clay plaster. In one of the fragments of a flute-like nozzle there was an orifice; similar to those found at "Tsetskhlauri I" and "Askana IV" (see above). At the bottom of the furnace was a layer of crushed charcoal heavily smeared with clay.

At a distance of 70-80 cm north of the furnace there was a 300×130-cm strip burnt red, most probably the base of the composite stone table (anvil) meant for working the spongy mass with a beetle.

The waste dump was partly on the southern slope of the hill where, despite repeated efforts, neither tea bushes, nor tung trees, would take root. Excavation of the dump yielded a great many lumps of slag (about 4 m^3) with a thick grey patina and the clay plaster of the furnace, ash-grey in colour; here were also found fragments of nozzles for bellows (Fig. 38 nos. 3-5), a miner's stone hammer (Fig. 38 no. 7), stone mortars and pestles, stones from the structure of the furnace, potsherds, as well as ash and pieces of charcoal.

There are a number of indications of date. In particular, the ceramic material, the structure of the furnace and the aspect of the iron slag assign the workshop to the pre-Antique period. Investigation of samples of charcoal from the furnace and the pit for installing the bellows, carried out in the Dating Laboratory in Tbilisi State University yielded the following data:

1. Sample TB-401, taken from "Mziani II" iron-smelting furnace at a depth of 30 cm from the surface of the surviving subterranean part of the furnace showed the following: an absolute date 2525±50 = 575 B.C., which is, according to the scale of Daymon *et al.*, 2578±102 = 628 B.C.
2. Sample TB-402, taken from "Mziani II" iron-smelting furnace at a depth of 50 cm from the surface of the surviving subterranean part of the furnace showed the following: an absolute date 2890±50 = 940 B.C., which is, according to the scale of Daymon *et al.*, 3022±80 = 1072 B.C.
3. Sample TB-403, taken from "Mziani II" iron-smelting furnace at a depth of 120 cm (i.e. from the bottom of the furnace) showed the following: an absolute date 3250±50 = 1280 B.C.,[17] which is,

[17] A.A. Burchuladze, G.I. Togonidze. *op. cit.*, samples TB-401, 402, 403.

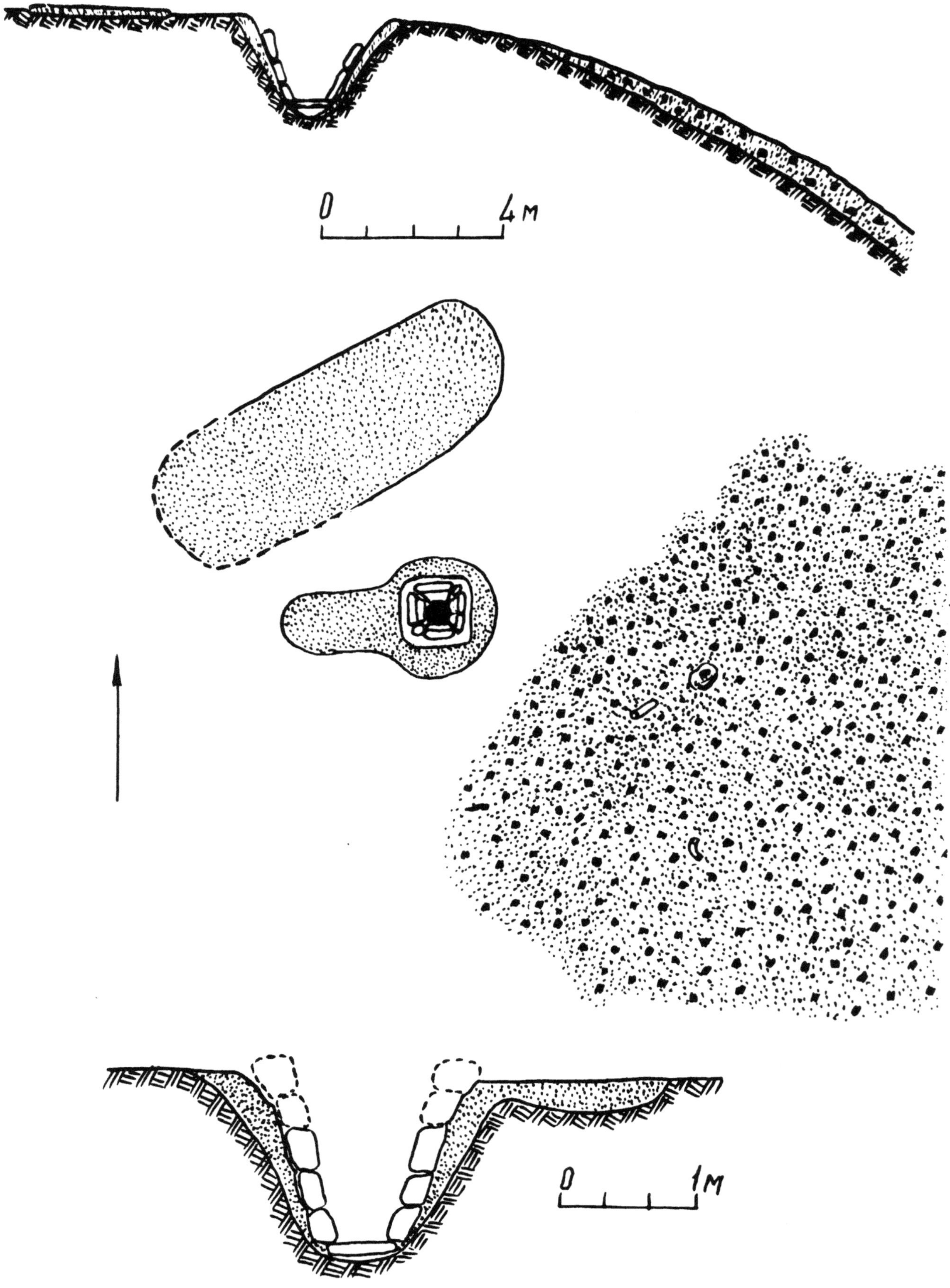

Fig. 37. Iron-smelting workshop "Mziani II." Ground plan and sections.

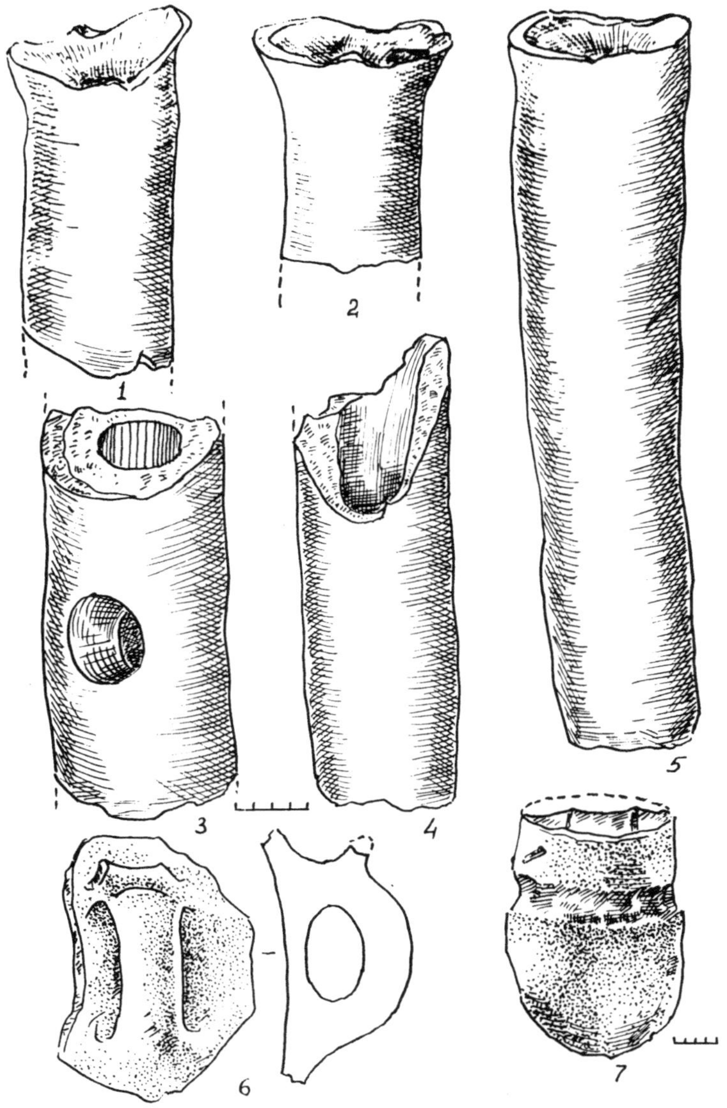

Fig. 38. "Mziani II." 1-5 – fragments of clay nozzles for bellows; 3 – fragment of a nozzle with a hole; 6 – fragment of a vessel with a zoomorphic handle; 7 – fragment of a stone mallet with a groove.

according to the scale of P.E. Daymon *et al.*, 3445±125 = 1495 B.C.[18]

Thus:

a) we have from the iron-smelting furnace "Mziani II" a series of samples taken from different levels of the furnace and giving different absolute chronological indices; a characteristic circumstance is that the numerical value of the absolute age increases with the depth from which the sample was extracted; this suggests that in some cases the sterility of the sample, and, naturally, the exactness of dating grows along with the depth of the samples location;

b) this workshop turned out to be one of the most ancient in the early group of the sites in this centre; from investigations carried out in recent years, "it appeared that in the period preceding the first millennium B.C. the concentration of carbon in the atmosphere of the Earth was higher than in the later times, and, naturally, samples subjected to laboratory analyses show "younger" dates because of the higher content of radioactive carbon."[19]

Concluding from the above, we may state that the "Mziani II" iron-smelting workshop functioned approximately in the 15th century B.C.

7. THE IRON-SMELTING WORKSHOP "MZIANI III" (excavated in 1983)

The remnants of the "Mziani III" iron-smelting workshops were identified by us in 1975 at the edge of

[18] Ch.A.Arslanov. *op. cit.*, p. 1161, pl.2.

[19] G. Kavtaradze. *On the Chronology of the Neolithic and the Bronze Age in Georgia.* Tbilisi, 1983, p. 6

the tea plantation, north of farmer Ipolité Kiguradze's homestead and east of farmer Elgudja Dolidze's holding, on a plot with a southern exposure and sloping towards the Kurepa river. On the north the plot is limited by a deep dry gully, wooded on both sides.

Three waste dumps were identified on this plot at a distance of a few metres from each other, two of them in the northern portion of the plot, and the third at the edge of the tea plantation (Fig. 39). The first two are partly overgrown with vegetation, but waste materials are noticeable here and there, especially in the middle of the dumps. The third dump is situated in the south-western part of the plot, at the edge of the plantation and is free of vegetation which could not develop here because the cultural level was saturated with lumps of slag. We should note here that the local residents battled against waste dumps, injuring them badly. Having noticed that ancient metallurgists used to store highly malleable fire-clay beneath the dumps, the local people made use of these stores of clay to make lids for hermetically sealing their large wine-jars ("kvevri," "karasi") buried in the ground to preserve their wine. Consequently, the waste dumps of "Mziani III" were badly damaged.

One of the workshops presented remnants of the iron-smelting furnace, the composite table (anvil) and a waste dump in the east part of the plot (Fig. 40), while its main waste dump was mostly on the north-east slope of the gully.

The remnants of the furnace and the composite table were covered with a layer of soil and clay 30-40 cm thick.

The iron-smelting furnace "Mziani III" has survived as a pit shaped like an upturned truncated pyramid, wholly faced with roughly hewn stone slabs of sandstone (Fig. 40). The bottom of the pit was a single flat cobblestone. Inside, the furnace was plastered with fire-clay, fragments of which remained in the south-eastern corner of the furnace. The soil around the furnace was burnt red by the intense heat, from a distance of 15-20 cm.

Inside, the furnace was filled with scorched stones, lumps of iron slag and clay plaster, fragments of clay nozzles and potsherds (Fig. 41), as well as clay, ashes and crushed charcoal which was taken to be analyzed.

The dimensions of the furnace are: height 80 cm, width at the top 70-80 cm, in the middle 50 cm, at the bottom 30 cm.

The composite table (anvil) was found south of the furnace at a distance of 120 cm. It is oriented east-west (Fig. 40); the scorched strip measures 430×150 cm. Only part of the stone pavement survives. Above the damaged part of the table a charred board of unknown use was discovered, dated by radiocarbon analysis to come from 1700 B.C.[20] This circumstance, supported by other data, suggests that the charred board above the composite table had nothing to do with the smelting workshop.

The waste dump (20×10 metres) contained: a great many lumps of slag with a thick greyish patina (about 4 cubic metres), ash-grey fragments of nozzles for bellows and unidentifiable potsherds. All this was mixed with clay, a lot of ash and pieces of charcoal. In the middle the dump was about 30 cm thick.

We have at our disposal several indications of the date of the workshop: the structure of the furnace (Fig. 40) and the aspect of the slag point to the pre-Antique period; judging by the pottery, the workshop can be dated to the first quarter of the first millennium B.C., while the radiocarbon analysis of samples taken from inside the furnace shows the following:

1. Sample TB-404, taken from iron-smelting furnace "Mziani III, 1," at a depth of 80 cm from the surface of the surviving burried part of the furnace indicates an absolute age of 2535±50 = 585 B.C.[21] which, according to the table compiled by Daymon *et al.* corresponds to 2588±102 - 640 B.C.[22]

2. Sample TB-405 taken from the same furnace at a depth of 50 cm from the surface of the surviving burries part of the furnace indicates an absolute age to be 2850±900 B.C.,[23] which, according to the table compiled by P.E. Daymon *et al.* corresponds to 2970±80 = 1020 B.C.[24]

Thus, the iron-smelting workshop "Mziani III, 1" may be dated to the turn of the second and the first millennia B.C.

8. IRON-SMELTING WORKSHOP "MZIANI III, 2" (excavated in 1983)

According to old residents of the locality, this workshop, including the smelting furnace and the composite table (anvil), was destroyed when the plot of land was subjected to intense cultivation. The only part left was the waste dump, situated near the dump of the workshop "Mziani III, 1," to the west of it. Excavation of the dump yielded, besides a great number of lumps of iron slag with a thick patina and pieces of ash-grey clay plaster of the furnace, ash and pieces, of charcoal, also some most interesting fragments of clay nozzles for bellows (Fig. 42), including two of the type directly connected with the leather "nipple" of the bellows (Fig. 42, nos. 2, 3). Such a clay nozzle we found, as mentioned above, in 1975 during the excavation of the waste dump of the iron-smelting workshop "Askana III, 1."[25] Until quite recently

[20] Burchuladze and Togonidze. *op. cit.*, TB-406.

[21] Burchuladze and Togonidze. *op. cit.*, TB-104.

[22] Arslanov. *op. cit.*, p. 1165, pl. 2.

[23] Burchuladze and Togonidze. *op. cit.*, TB-405.

[24] Arslanov, *op. cit.*, p. 1161, pl. 2.

[25] D.A. Khakhutaishvili. *Production Centre …*, pp. 23-24.

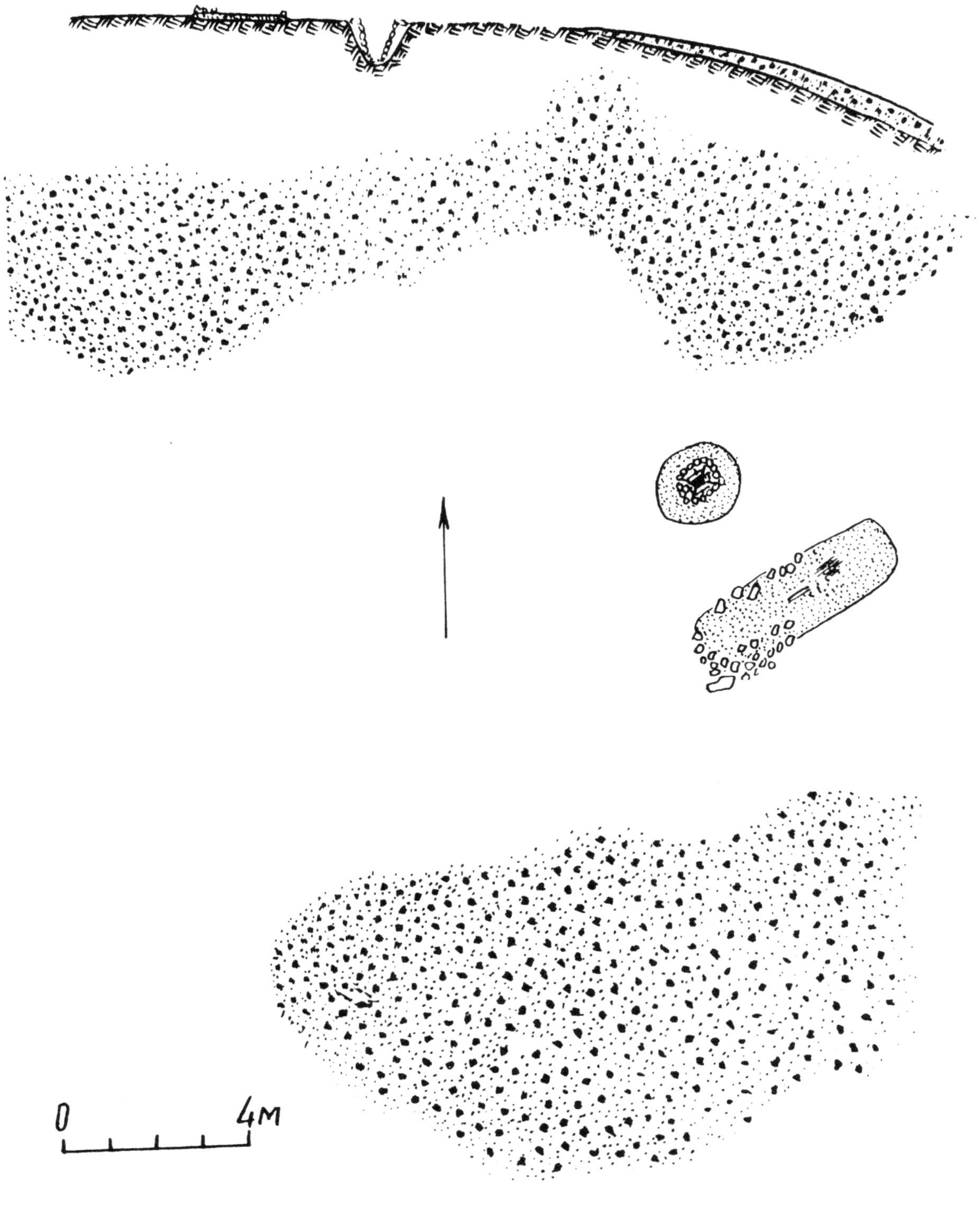

Fig. 39. Iron-smelting workshop "Mziani III." Ground plan and section.

Georgian folk metallurgy preserved this method of joining the leather and clay parts of bellows.[26]

Contemporary Georgian folk culture has, incidentally, preserved quite a number of objects of household and everyday use that go back to the Bronze Age (the hoe, the axe, the adze, the sickle, the threshing board, kitchen pottery, etc.).

[26] N.Rekhviashvili. *Folk Metallurgy of Georgia*. Tbilisi, 1964, pp. 92-96 (in Georgian, conclusions in Russian).

Thus, the continuity of metallurgical traditions can be followed through the course of several millennia.

8. IRON-SMELTING WORKSHOP "MZIANI III, 3" (excavated in 1983)

Nothing remains of this workshop except the waste dump, and even that has not fared well: the local residents found a considerable store of highly plastic, well-refined clay and had dug almost the whole of the dump already

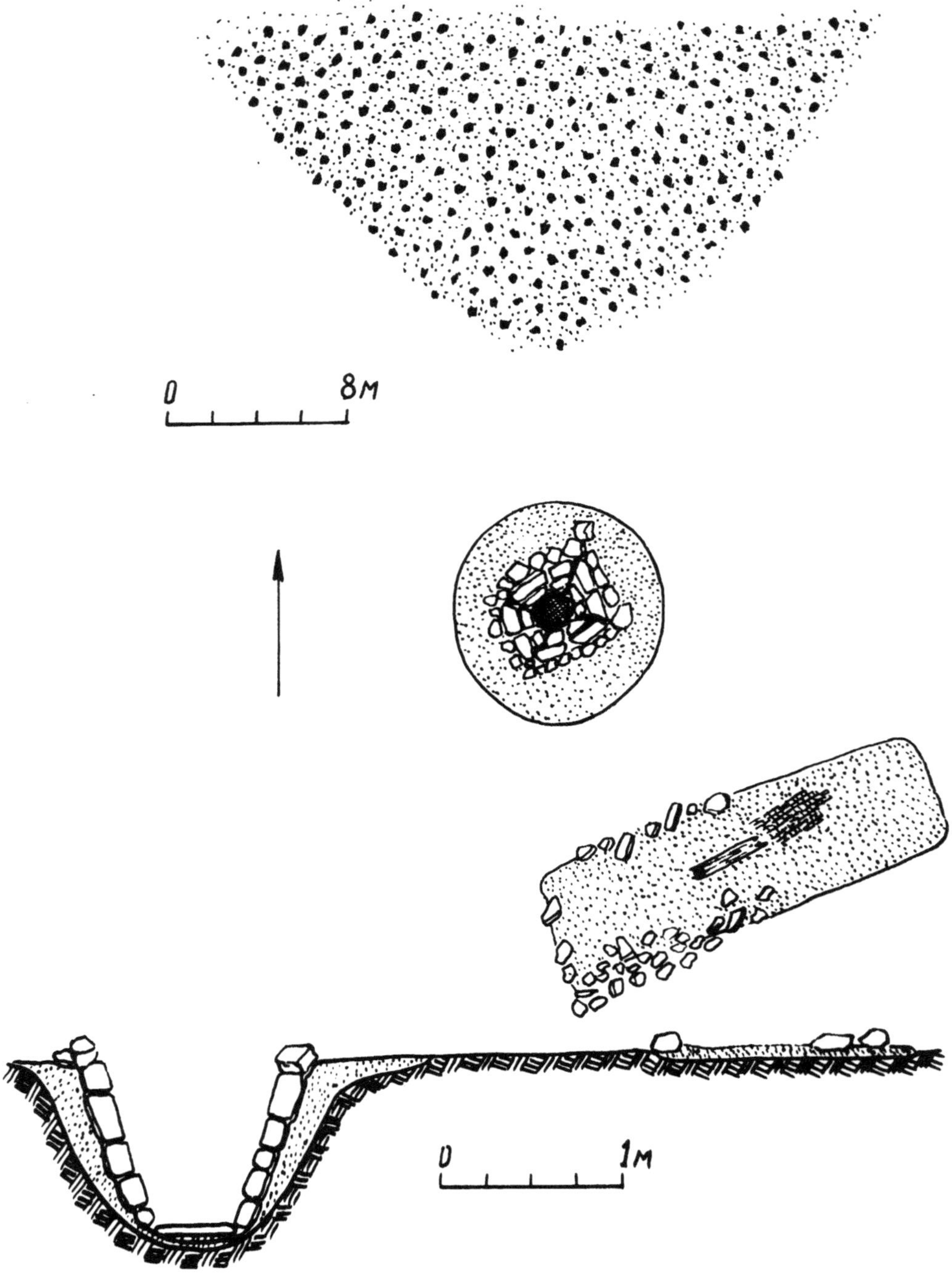

Fig. 40. Iron-smelting workshop "Mziani III, 1." Ground plan and section.

damaged by the former owner of the land to get at it. The dump contained lumps of iron slag with a thick grey patina and ash-grey plaster of the furnace, fragments of nozzles for bellows and potsherds (Fig. 43). All this was mixed with clay, ash and crushed charcoal.

The pottery found in this waste dump is closely analogous to the pottery from sites unearthed on the New Black Sea littoral terrace in Kobuleti, Ureki, Bobokvati and Chakvi. These data assign "Mziani III, 3" to the eighth-seventh centuries B.C.

9. IRON-SMELTING WORKSHOP "MZIANI IV" (excavated in 1983)

The iron-smelting workshop "Mziani IV" was discovered in 1975, west of farmer Amiran Dolidze's holding, at a distance of 60 metres from the edge of the lot, on a slope facing south bounded by a nameless gully overgrown with tall trees. In the middle part of the slope a flat terrace-like area had been made bordered on the south by a waste dump covering about 100 square metres.

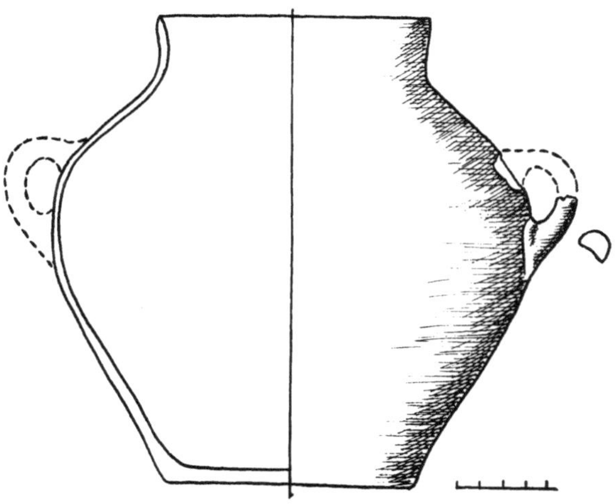

Fig. 41. "Mziani III, 1." Vessel with two handles.

The workshop consisted of the remains of an iron-smelting furnace, a composite table (anvil) and a waste dump (Fig. 44).

The iron-smelting furnace "Mziani IV" was identified on a plot beneath a sterile layer of yellow diluvial clay 30 cm thick. It has survived as a pit shaped like an upturned truncated pyramid, faced with stone from top to bottom (the top courses had collapsed); the floor of the furnace was a flat cobble stone. On the facing stones were traces of fire-clay plaster, ash-grey in colour.

The height of the surviving part of the furnace is 130 cm; the width: at the top (as we suppose) 80 cm, in the middle 65 cm, and at the bottom 30 cm.

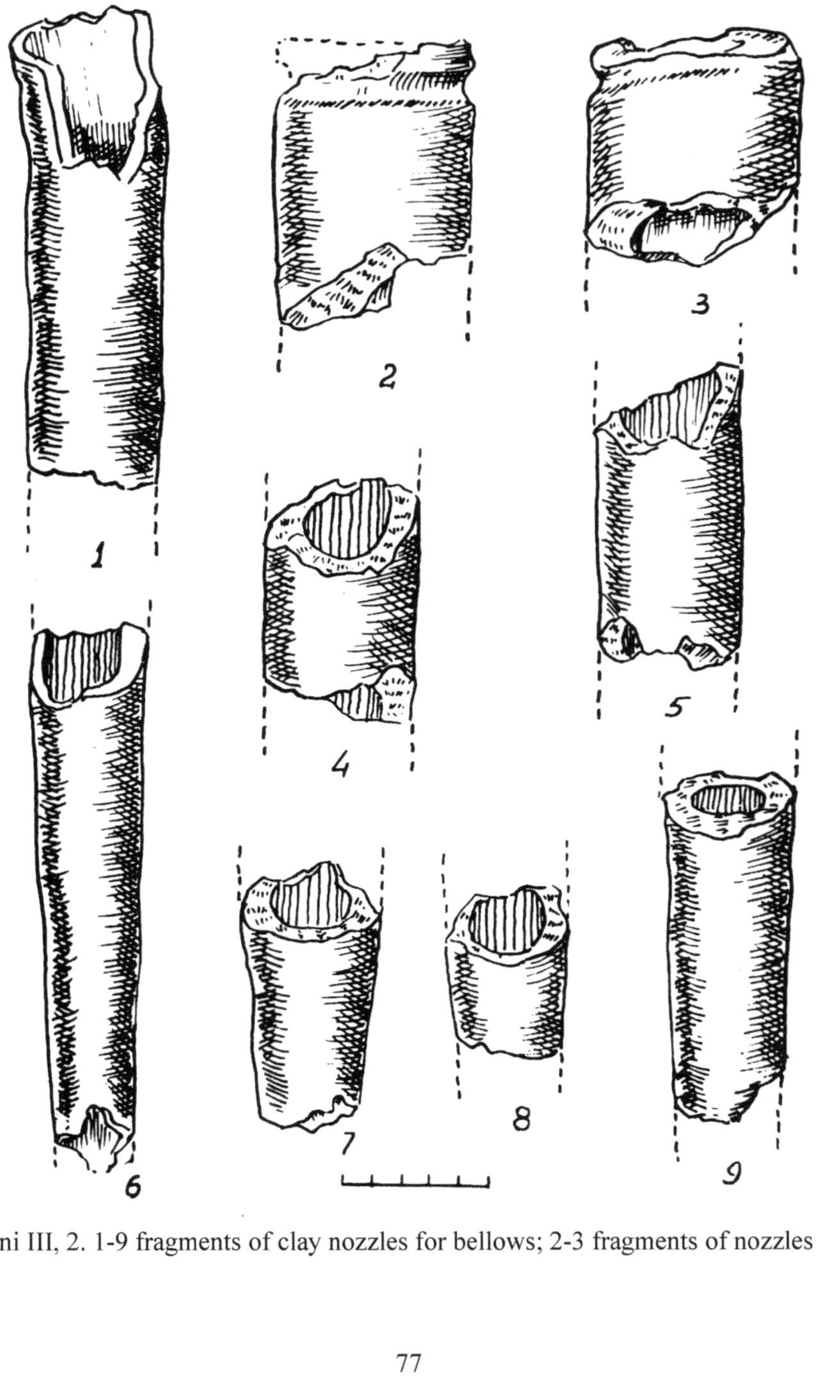

Fig. 42. "Mziani III, 2. 1-9 fragments of clay nozzles for bellows; 2-3 fragments of nozzles with a groove.

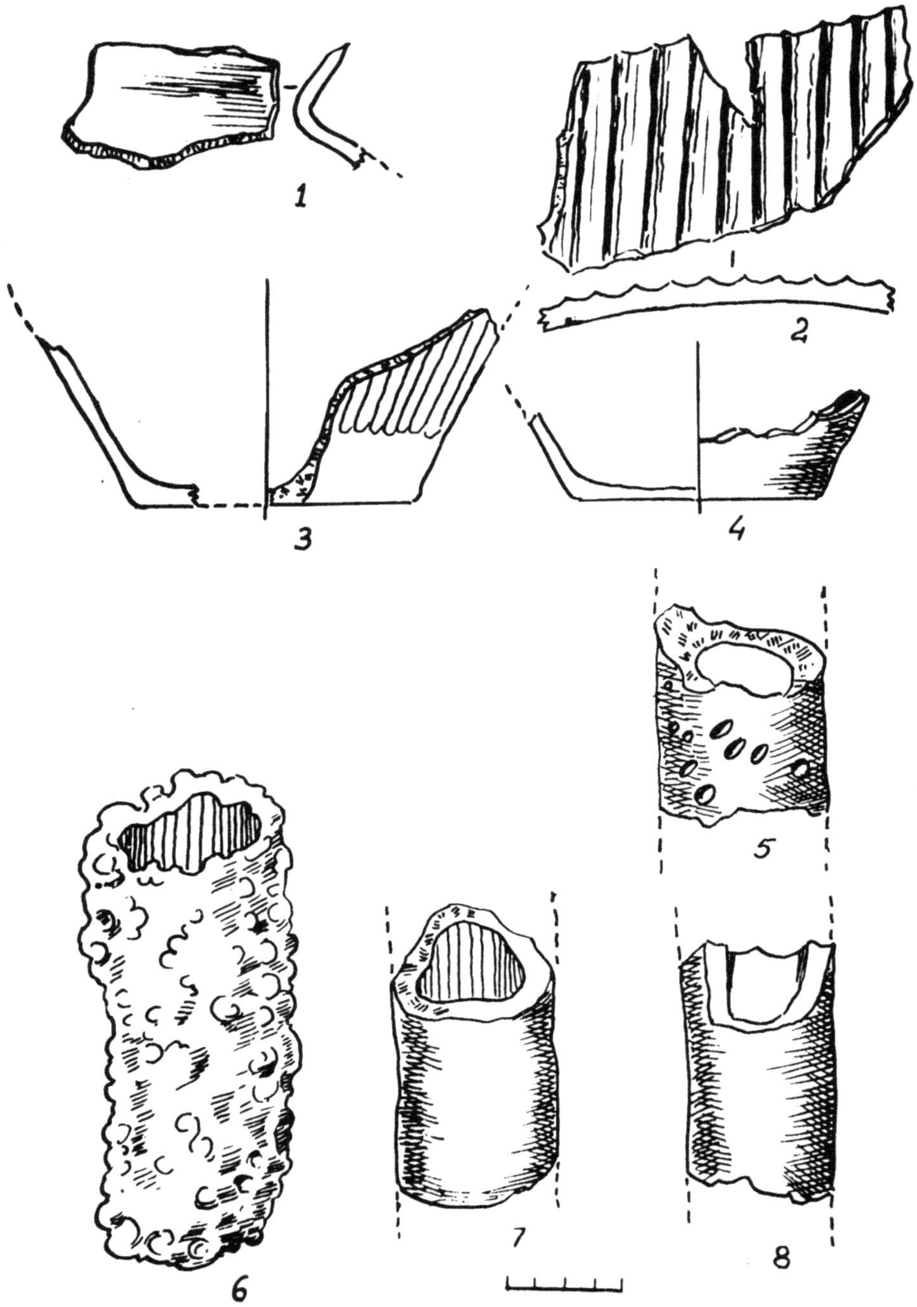

Fig. 43. "Mziani III, 3." 1 fragment of a pot; 2-3 fragment of a thick-walled burnished clay vessels with broad fluting; 4 bottom of a vessel; 6-8 fragments of clay nozzles for bellows.

The inner space of the smelting furnace was filled with scorched stones from the structure of the furnace, lumps of iron slag and pieces of clay nozzles for bellows (Figs. 45, 1-3), fragments of clay plaster and potsherds (Fig. 45. 4-6), ash and pieces of charcoal. All this was mixed with yellow clay. The earth around the pit of the furnace was burnt red by the heat for a distance of 15-20 cm. On the south-western side the bellows was installed, the marks of which remained as a pit 50 cm in diameter and 40 cm deep. It was filled with a mixture of clay and crushed charcoal. The edge of the pit was burnt reddish by the heat.

Of the composite table (anvil) there remained the base, a strip of diluvial clay, 500×200 cm, burnt red. The table was at a distance of 150 cm (Fig. 44) to the north of the furnace. The stones that had formed the table were scattered all over the hill.

The waste dump, located north of the furnace on a steep slope was badly damaged by people hunting for tempered malleable clay used for hermetically sealing the mouths of earthenware wine-jars. A considerable part of the waste must have spilled down the steep slope to the bottom of the gully and had been washed down by the stream.

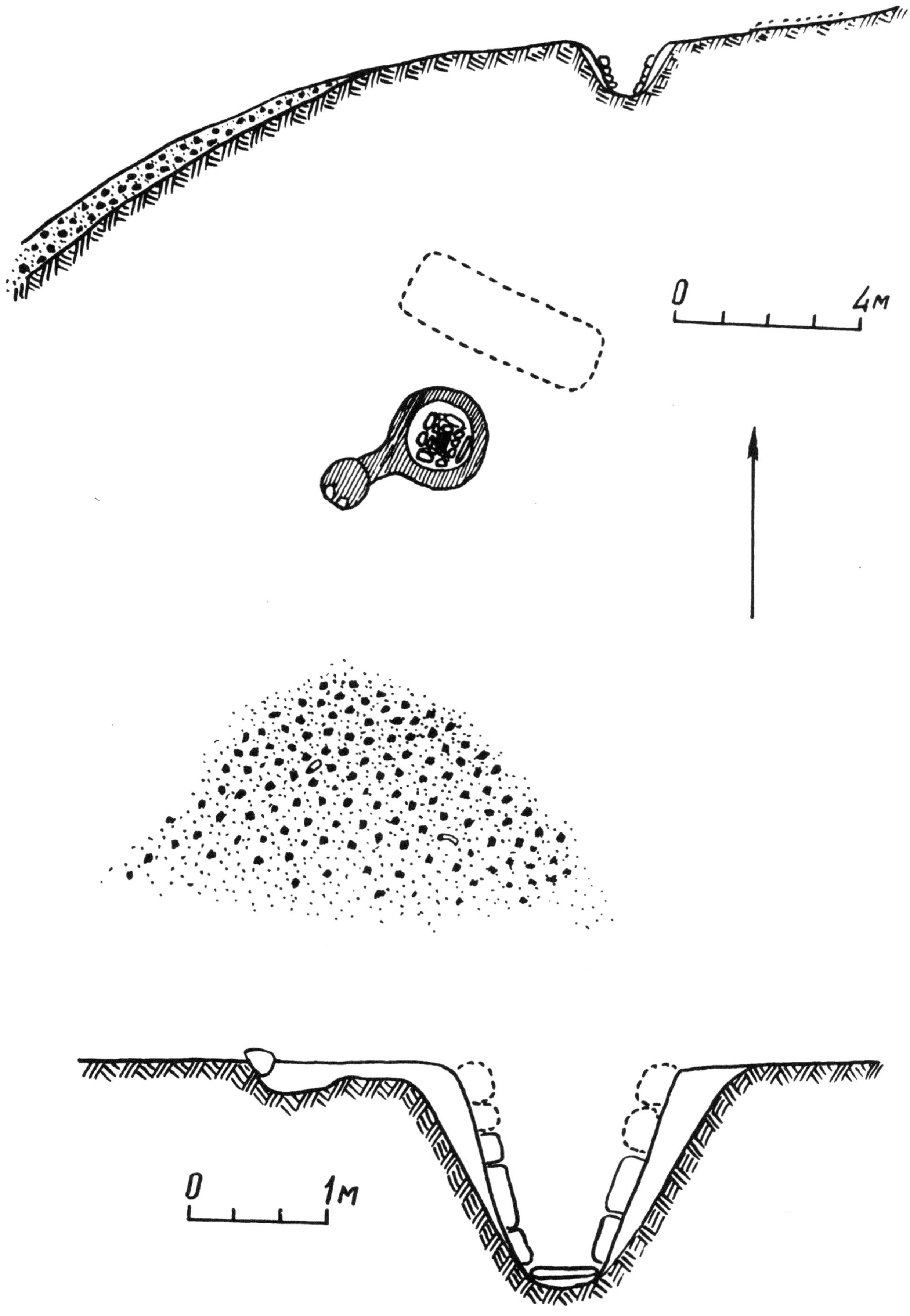

Fig. 44. Iron-smelting workshop "Mziani IV." Ground plan and sections.

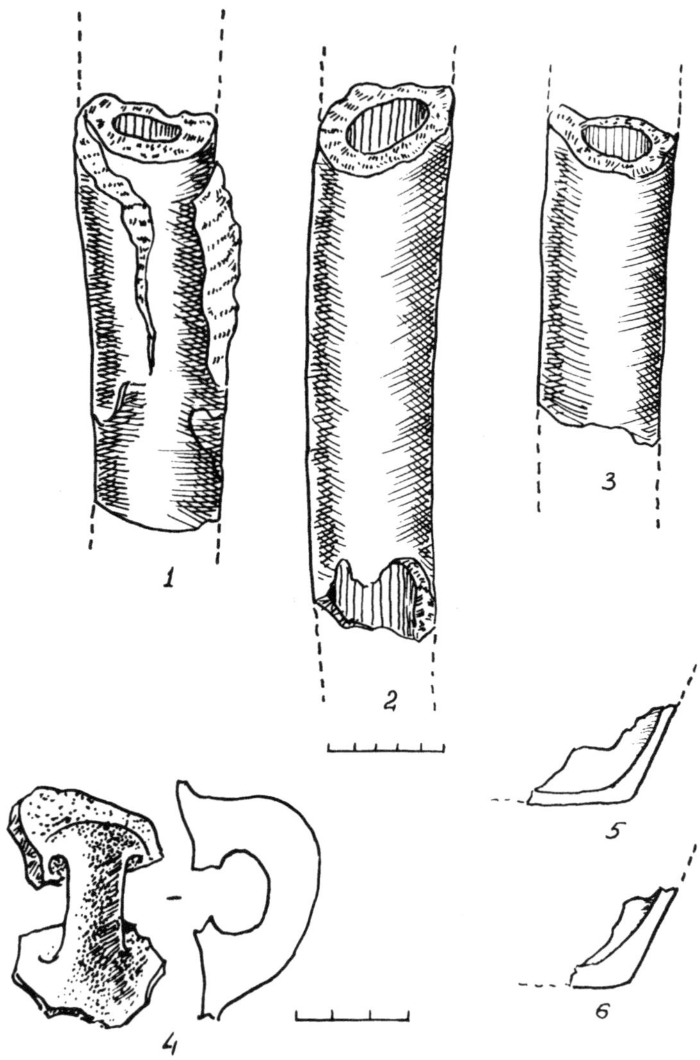

Fig. 45. "Mziani IV." 1-3 fragments of clay nozzles for bellows; 4 handle of a vessel; 5-6 fragments of clay bases of vessels.

Excavation of the remaining part of the dump yielded a considerable amount of iron slag (up to 2 cubic metres) with a thick greyish patina, fragments of clay plaster of the furnace, ash-grey in colour, fragments of nozzles for bellows (Fig. 45, 1-3) and non-diagnostic potsherds. Here there also occurred scorched stones from the structure of the furnace and the table, as well as ashes and pieces of charcoal. Around the waste dump there had sprung up a "plantation" of ferns, a plant characteristic of the entire area with diluvial yellow clay. These plants have a powerful root system, in places reaching three or more metres deep.

Potsherds (Fig. 45, 4-6) are no indication as to the age of this workshop as they were found away from the furnace, in the area around it. Unfortunately, the sample of charcoal taken from inside the furnace (sample TB-409) could not be analyzed because it was badly contaminated with various admixtures and by clay. To the joy of the archaeologists, a small piece of charcoal taken from the bellows pit turned out to be absolutely sterile. The absolute date of this sample (TB-408), according to the Dating Laboratory of Tbilisi State University is determined as 3170±45 = 1220 B.C.,[27] which, according to the table compiled by Daymon *et al.* corresponds to 3365±125 = 1415 B.C.[28]

Thus, according to the date of radiocarbon analysis, the iron-smelting furnace "Mziani IV" functioned at the end of the 15th or at the beginning of the 14th century B.C. There are no grounds for questioning the suggested dating.

[27] Burchuladze and Togonidze. *op. cit.*, sample TB-408.

[28] Arslanov. *op. cit.*, p. 1161, pl.2.

IRON-SMELTING WORKSHOPS OF THE "MSHVIDOBAURI" AND "NAGOMARI" GROUPS

10. IRON-SMELTING WORKSHOP "MSHVIDOBAURI I" (excavated in 1984)

The iron-smelting workshop "Mshvidobauri I" is situated west of the tea factory in the village of Askana, at a distance of 100 metres from the main building of the enterprise. The existence of an iron-smelting shop was indicated by a large waste dump devoid of any vegetation on a slope with a north-western exposure. The site was identified in 1975 in the gully of the Tskaltsitela I river, at the very top of the watershed elevation Tskaltsitela I and Tskaltsitela II.

The workshop has come down to us with the remains of an iron-smelting furnace, a composite table (anvil) and a waste dump (Fig. 46). Above the slope where the waste dump lies, there is an level space of plot about 150 m^2, the western part of which was overlain by a portion of the waste dump.

The iron-smelting furnace "Mshvidobauri I" was identified on a completely level plot after a 40-cm-thick sterile layer of yellow clay had been removed. The furnace was a pit shaped like an upturned truncated pyramid (Fig. 46) faced with stone. Only part of the facing stones have survived, and that only on the east wall of the furnace. The soil around the pit is burnt red, 15-20 cm thick.

The height of the surviving underground part of the furnace in 100 cm; the width: at the top about 80 cm, in the middle about 45 cm, at the bottom 30-35 cm.

Inside, the pit was filled with clay, scorched stones from the structure of the furnace, lumps of iron slag with a thick patina, greyish in colour and a mixture of crushed charcoal, ash and clay. Unfortunately, a sample of crushed charcoal taken from the bottom of the furnace did not lend itself to radiocarbon analysis (TB-431).

Of the composite table there remained a strip of clay burnt red, 20 cm east of the furnace. The table, which must have been 410x110 cm in size, judging by the size of the burnt strip, was most probably destroyed long ago (Fig. 40).

The waste dump, most of which (as mentioned above) was on the slope facing north-west, was free of vegetation. The area covered by it was about 400 square metres, the depth in the centre about 40 cm. In the upper part, under a layer of slag and ash, was a store of highly malleable, well-refined clay. The dump contained a great number of lumps of clay and slag, with the thick grey patina characteristic of pre-Antique slags, fragments of ash-grey plaster, from the walls of the furnace, and variously shaped stone mortars (Fig. 47).

The dump was full of ashes and pieces of charcoal, fragments of nozzles for bellows and potsherds. Among the latter, some are noteworthy: shreds of a thick-walled vessel adorned outside with broad fluting (Fig. 48) and the handle of a vessel with a roll and notches, a type widespread in the eighth-seventh centuries B.C. (Fig. 48, 3).

As for the dating of the "Mshvidobauri I" workshop, we might stress the structure of the furnace and the characteristic features of the iron slag and the clay plaster. Judging by the ceramic material, we may assign it to approximately tenth-eighth centuries B.C.: it was in this period that thick-walled vessels were produced in Colchis and were decorated with broad fluting.[29]

11. IRON-SMELTING FURNACE "MSHVIDOBAURI II" (excavated in 1984)

The iron-smelting workshop "Mshvidobauri II" is situated at a distance of 109 metres from "Mshvidobauri I," to the southwest of the latter. It occupies the south- and west-oriented slopes of a promontory-like hill on which small waste dumps are discernible. On two sides, the hill is bordered by dry gullies which come together at the "nose" of the spur.

The remains of the workshop consist of: the ruins of the iron-smelting furnace, of the composite table (anvil) and the waste dump (Fig. 49). What was left of the iron-smelting furnace was discovered at the top of the hill beneath a 30-40 cm thick layer of humus and yellow clay. The furnace has survived as a pit, shaped like an upturned truncated pyramid. The upper three-fifths of the furnace had been faced with stones, the lower hemispherical part was plastered with a thick layer of fire-clay.

The dimensions of the surviving part of the furnace are: height 125 cm, width at the top 85 cm, in the middle 45 cm, at the bottom 30 cm.

The interior of the pit was filled with scorched stones from the structure of the furnace, lumps of slag, fragments of clay plaster and nozzles for bellows, ashes and a mixture of clay and charcoal. Lower down were found some non-diagnostic potsherds.

What remained of the composite table (anvil) was found east of the furnace, 130 cm away. The base of the table, burnt red, was 140 cm long by 90 cm wide.

The waste dump lying on both slopes of the hill was 30 cm deep and contained, mainly, scorched stones from the structure of the furnace and the composite table, lumps of iron slag with a thick grey patina, fragments of ash-grey plaster the furnace, fragments of nozzles for bellows and pieces of charcoal.

---

[29] Khakhutaishvili. *On the Chronology...*, pp. 136-139.

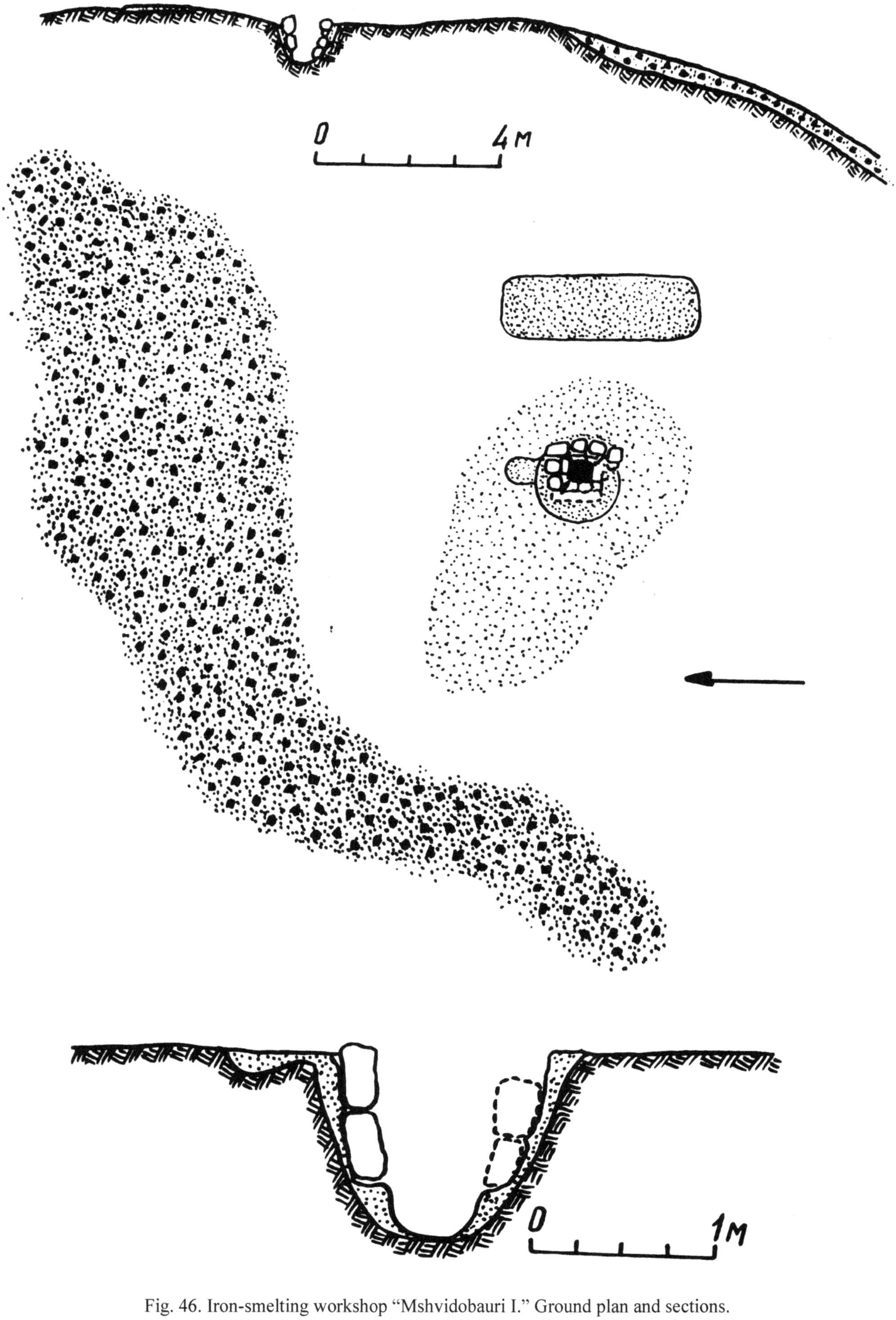

Fig. 46. Iron-smelting workshop "Mshvidobauri I." Ground plan and sections.

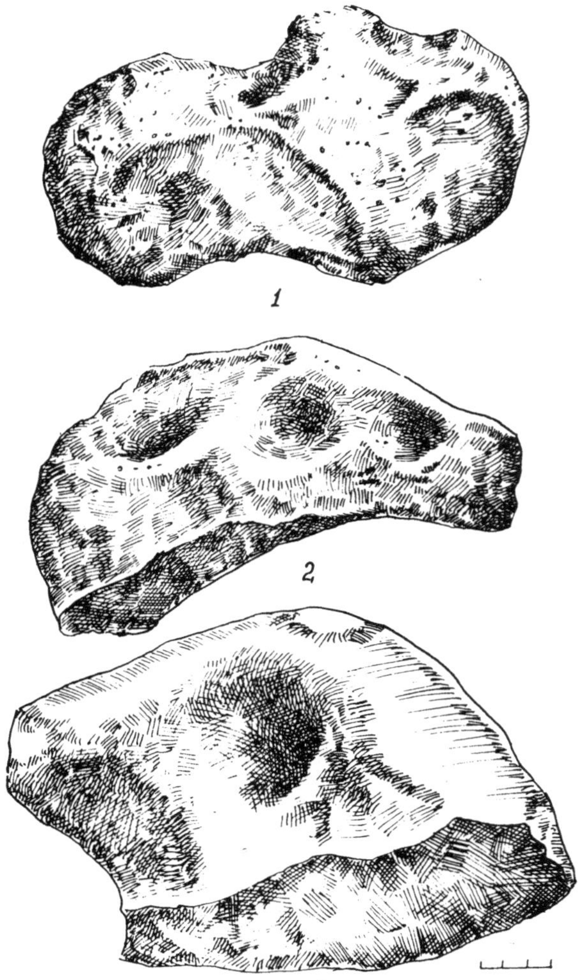

1

2

3

Fig. 47. "Mshvidobauri I." 1-3 – fragments of mortars for crushing flux.

Judging by its construction, as well as the characteristic peculiarities of the slag and plaster, the workshop undoubtedly functioned in the pre-Antique period. Unfortunately, but little information for dating can be gleaned from the fragments of ceramic vessels; however, the results of radiocarbon analysis of samples of charcoal taken from the furnace were most valuable:

1. Sample TB-432. From "Mshvidobauri II" iron-smelting furnace gave an absolute date of 3483±125 = 1533 B.C.,[30] which, according to the table compiled by Daymon *et al.* corresponds to 3760±103 = 1810 B.C.;[31]
2. Sample TB-433. From "Mshvidobauri II" iron-smelting furnace gave an absolute date of 3270±50 = 1320 B.C.,[32] which, according to Daymon *et al.* corresponds to 3490±125 = 1540 B.C.[33]

Thus, the iron-smelting workshop "Mshvidobauri II" is a most interesting centre of ancient iron production.

[30] Written certificate by Burchuladze and Togonidze from the Dating Laboratory, dated 1985, Tbilisi University.
[31] Arslanov, *op. cit.*, p. 1161, pl.2.
[32] Written certificate by Burchuladze and Togonidze.
[33] Arslanov. *op. cit.*, p. 1165, pl.2.

We hope that further investigation will clarify the puzzling situation in which specialists who study the history of Colchian iron metallurgy find themselves at present.

12. IRON-SMELTING WORKSHOP "MSHVIDOBAURI III" (excavated in 1984)

The remnants of iron-smelting shop "Mshvidobauri III" were identified 300 metres south of "Mshvidobauri II," on an elongated spur with two slopes. Along the ridge of the spur there is a road running from east to south; the furnace and the composite table (anvil) must have been destroyed when the road was built. A search for these main elements of an ancient iron-smelting workshop proved fruitless.

The waste dump was spread over both slopes of the spur, but its main part turned out to be on the south slope, as further excavations proved, close to the road.

The dump contained a large amount of iron slag with a thick grey patina and pieces from fire-clay plaster of the furnace, ash-grey in colour, scorched stones from the structure of the furnace and from the table, several non-diagnostic fragments of nozzles of bellows and small pieces of charcoal.

Regarding the age of this workshop, the only thing we can say is that it functioned in the pre-Antique period. We have no data with which to assign a more precise date.

Thus, in the vicinity of the Askana village tea factory (at Dabali Etseri) three sites were excavated in 1984, one of which turned out to be most interesting.

13. IRON-SMELTING WORKSHOP MSHVIDOBAURI IV" (excavated in 1984)

The remnants of "Mshvidobauri IV" iron-smelting workshop were identified near the left bank of the Tskaltsitela I brook, on a road leading to a plantation, at a place called "boga" (wooden bridge), to the north of Askana tea factory, 1000-1200 metres away from it. During archaeological survey of the region in 1975, local residents told us that a lot of waste material had been taken away to cover plantation roads in lieu of gravel. Indeed, the waste dump looked imposing despite the ravages that preceded our arrival, and it covered a large area. The south end of the dump descended as far as the bed of the stream, and the dump as a whole covered over 400 square metres. The north-western part of the dump bordered on a two-humped hillock with parts evidently levelled, where, as we supposed, the remains of at least two iron-smelting furnaces must have been. Above this elevation is the Mshvidobauri village tea plantation.

Archaeological excavation of the site proved that not only the waste dumps were been damaged, but so too had been

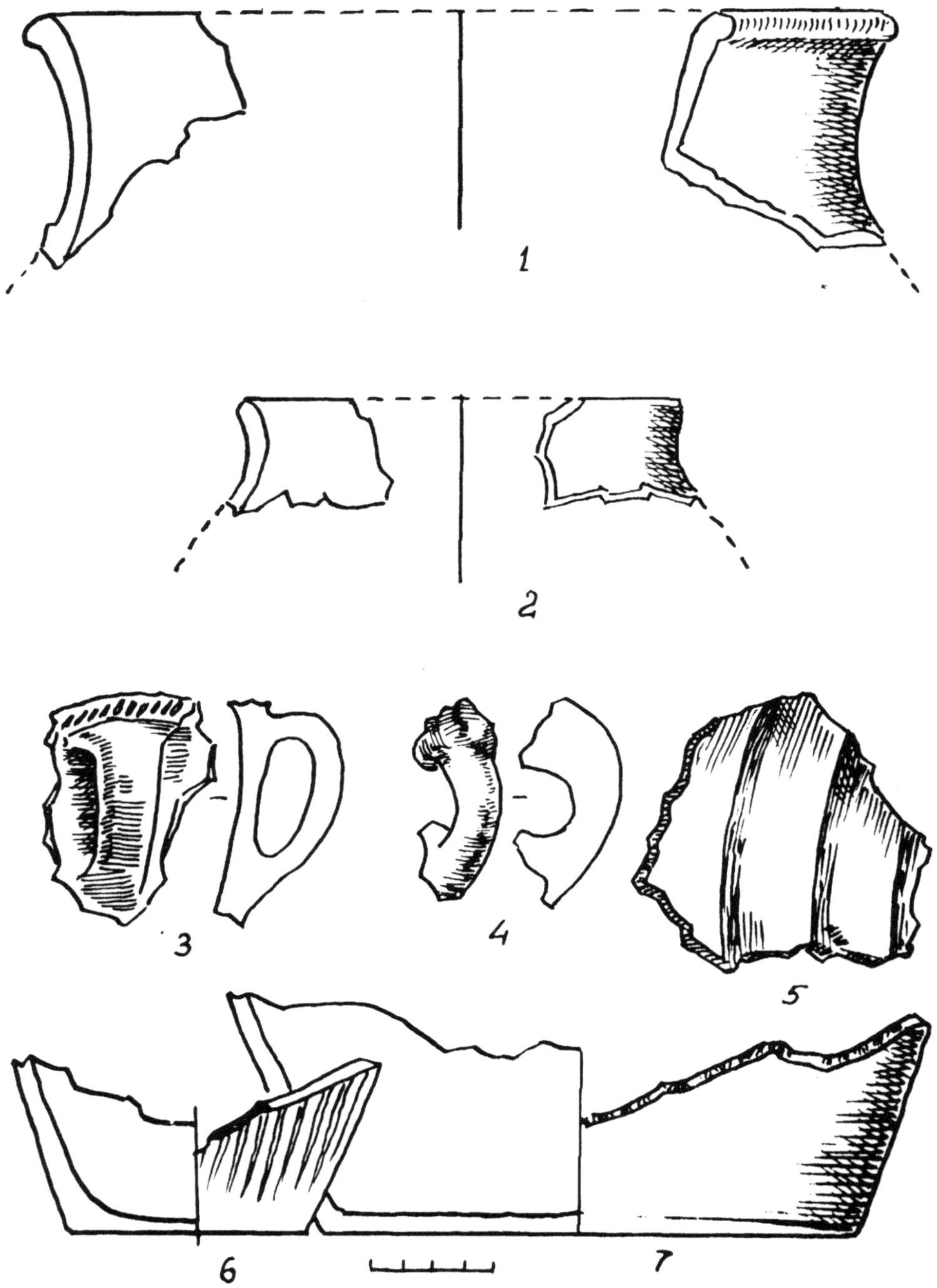

Fig. 48. "Mshvidobauri I." 1 – fragment of a jug; 2 – fragment of a pot; 3-4 – handles of vessels; 5 – fragment of a thick-walled burnished clay vessel with broad fluting; 6 – fragment of a vessel with narrow fluting; 7 – fragment of a vessel.

the other basic elements of iron-smelting workshops: the iron-smelting furnaces and the composite tables (anvils) (Fig. 50).

Only a part of the iron-smelting furnace "Mshvidobauri I" had survived. The furnace was built in a pit shaped like an upturned truncated pyramid. Clearing showed that only separate fragments of the stone facing remained in situ.

The surviving part of the furnace had the following dimensions: height 70 cm, width at the top 75 cm, at the bottom 30 cm.

Inside, what remained of the furnace was filled with lumps of slag, scorched stones from the structure of the furnace, fragments of nozzles for bellows, ash, clay and crushed charcoal. The south wall of the furnace was damaged by a pit that had been dug later almost down tothe lower course of the facing stones. At the bottom was a layer of crushed charcoal mixed with yellow clay and ashes.

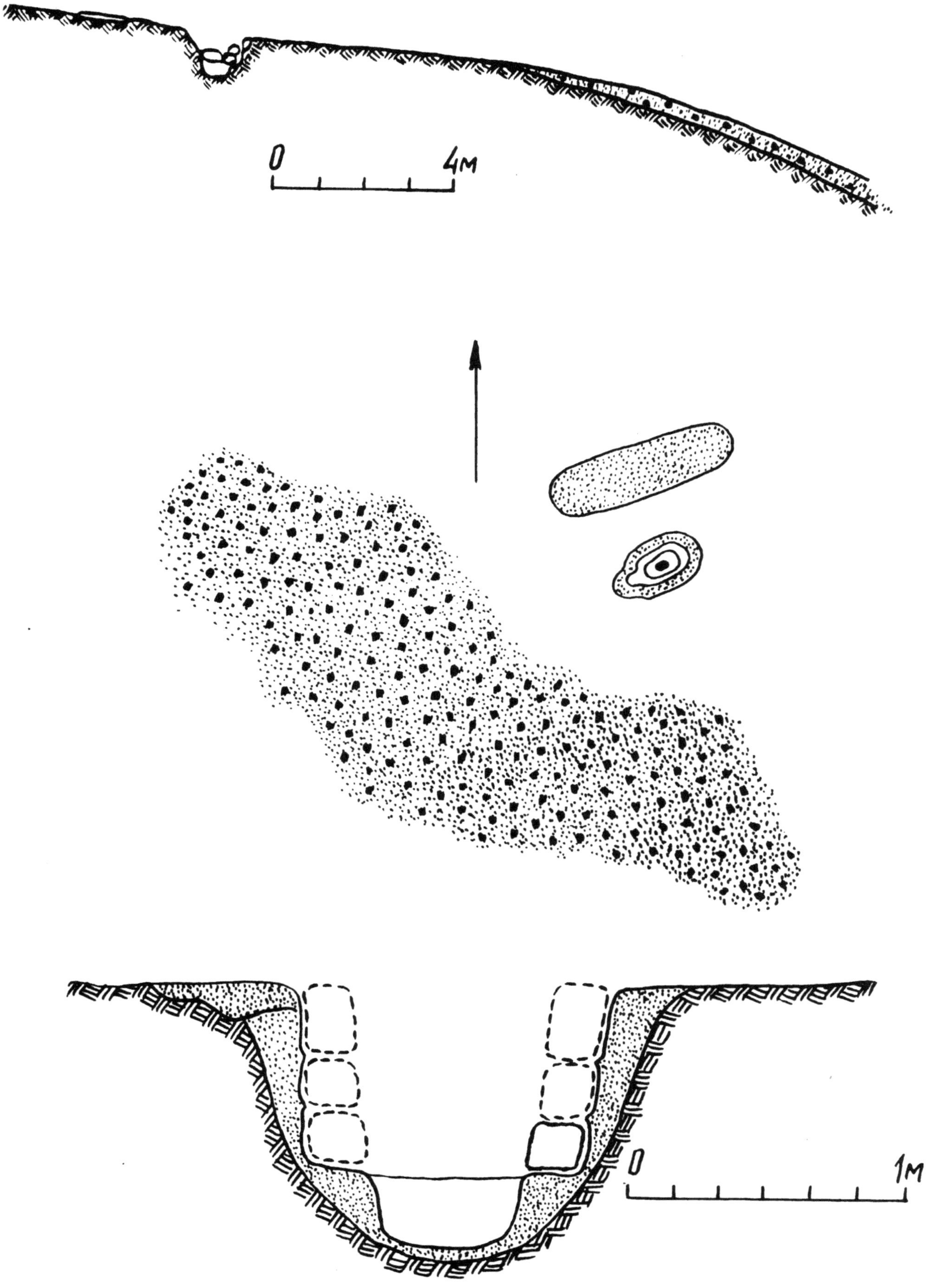

Fig. 49. Iron-smelting workshop "Mshvidobauri II." Ground plan and sections.

Remnants of the composite table (anvil), a strip of burnt clay 200×80 cm were identified north of the furnace, 160 cm away from it.

A furnace which belonged to another workshop has not survived. It must have been destroyed when the tea plantation was created or when the waste dump was removed with the aid of modern machinery.

Excavation of the remaining part of the waste dump unearthed a great many lumps of iron slag (about 10 cubic metres) with a thick greyish patina and pieces of fire-clay plaster, ash-grey in colour (about 4 cubic metres), fragments of nozzles for bellows (Fig. 51), ash, pieces of charcoal and numerous scorched stones from the
structure of the furnace and from the composite table, potsherds of various ages (Fig. 52), including zoomorphic handles (Fig. 52, 4, 6).

As indications for dating the "Mshvidobauri IV" workshops, we might mention the structure of the furnace, the peculiarities of the iron slag and the fire-clay plaster. These indices assign the above workshops to the pre-Antique period,

Analysis of the crushed charcoal taken from the bottom of the furnace and carried out at the Dating Laboratory of Tbilisi State University determined the absolute age of the sample as 3190±45 = 1240 B.C.,[34] which corresponds to 3390±1440 B.C.[35] according to Daymon *et al.*

IRON-SMELTING WORKSHOPS "NAGOMARI I"

As was ascertained in 1975, this archaeological site consisted of two dumps for waste materials and, consequently, remains of two iron-smelting workshops which, judging by the waste, functioned in the pre-Antique period. The site is situated on the ridge of a spur of a system of hills descending from north to south to the gorge of the Tskaltsitela II river at the junction of the lands belonging to the Mshvidobauri and Nagomari villages, at the edge of a tea plantation. One waste dump is on a west-facing slope, the other on a slope with an eastern exposure. On the ridge of the spur, between the waste dumps, there is a levelled area about 100 metres square, along whose centre there runs a plantation road which has partially damaged the dumps east of it. To the west, the site is bordered by a dry gully, to the east it follows a gentle slope planted to tea-shrubs.

The soils here are the same as in the entire foothill zone of the Eastern Black Sea littoral (see above): diluvial yellow clay covers thick strata of laterites formed by the erosion of Tertiary tufogenic rock. Here and there the laterites are denuded, forming an striking contrast of soils.

The central part of the waste dumps is, as a rule, devoid of vegetation. The cause of this has already been mentioned.

When the layers of humus and yellow clay had been removed, the contours of two iron-smelting furnaces became visible.

14. IRON-SMELTING WORKSHOP "NAGOMARI I,1" (excavated in 1984)

The iron-smelting workshop "Nagomari I,1" has survived with its essential parts: remains of the iron-smelting furnace, a composite table (anvil) and a waste dump on the west side of the hill (Fig. 53).

The smelting furnace "Nagomari I, 1" has survived as a pit faced with rough-hewn stones; in shape of an upturned truncated pyramid (Fig. 53). The bottom of the pit is a hemispherical hollow plastered with a thick layer of fire-clay. On the facing stones, are occasional fused purplish patches.

The height of the surviving part of the furnace is 100 cm, the width at the top 55-60 cm, in the middle 30-35 cm, at the bottom 20-25 cm.

South of the furnace, and almost adjacent to it, is a small pit where the bellows must have been installed.

Inside, the furnace was filled with clay, scorched stones from the structure of the furnace itself, lumps of slag, fragments of clay nozzles for the bellows, and pieces of clay plaster from the furnace, ash, and crushed charcoal mixed with clay and ash.

The composite table (anvil) oriented from west to east is to the north of the furnace, at a distance of 220 cm. It is made of flat stones (its western part), as well as fragments of clay plaster from the furnace; it covers an area 420×110 cm (Fig. 53).

The waste dump contained a lot of lumps of iron slag (about 5 cubic metres) with a greyish patina and fragments of clay plaster from the walls of the furnace, pestles (Fig. 56), fragments of nozzles for bellows (Fig. 54), ashes, pieces of charcoal, potsherds (Fig. 55), as well as scorched stones from the structure of the furnace and the composite table (anvil), evidently from another workshop. The dump covered about 200 square metres, but its depth in the centre did not exceed 25 cm.

The structure of the furnace, the external aspect of the slag and the clay plaster assign the workshop to the pre-Antique period. The analysis of the charcoal (crushed) carried out in the Dating Laboratory, Tbilisi State University, by Burchuladze and Togonidze proved that the "Nagomari I, 1" workshop functioned about 830 B.C. (2780±35) which corresponds to 940 B.C. (2890±80) according to Daymon *et al.* Concluding from this, we

[34] A written communication by Burchuladze and Togonidze from the Tbilisi University Dating Laboratory.
[35] Arslanov. *op. cit.*, p. 1161, pl. 2.

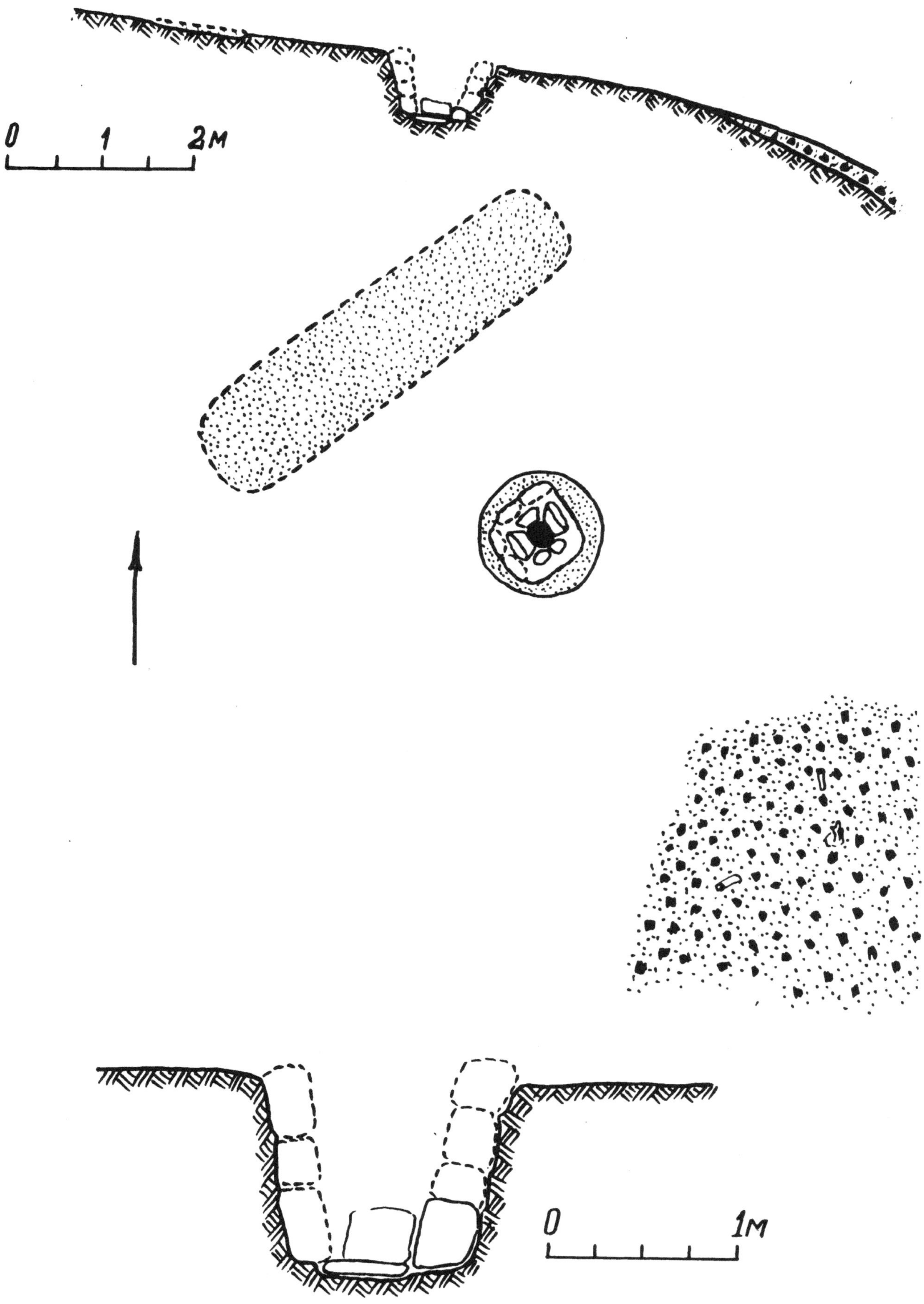

Fig. 50. Iron-smelting workshop “Mshvidobauri IV.” Ground plan and sections.

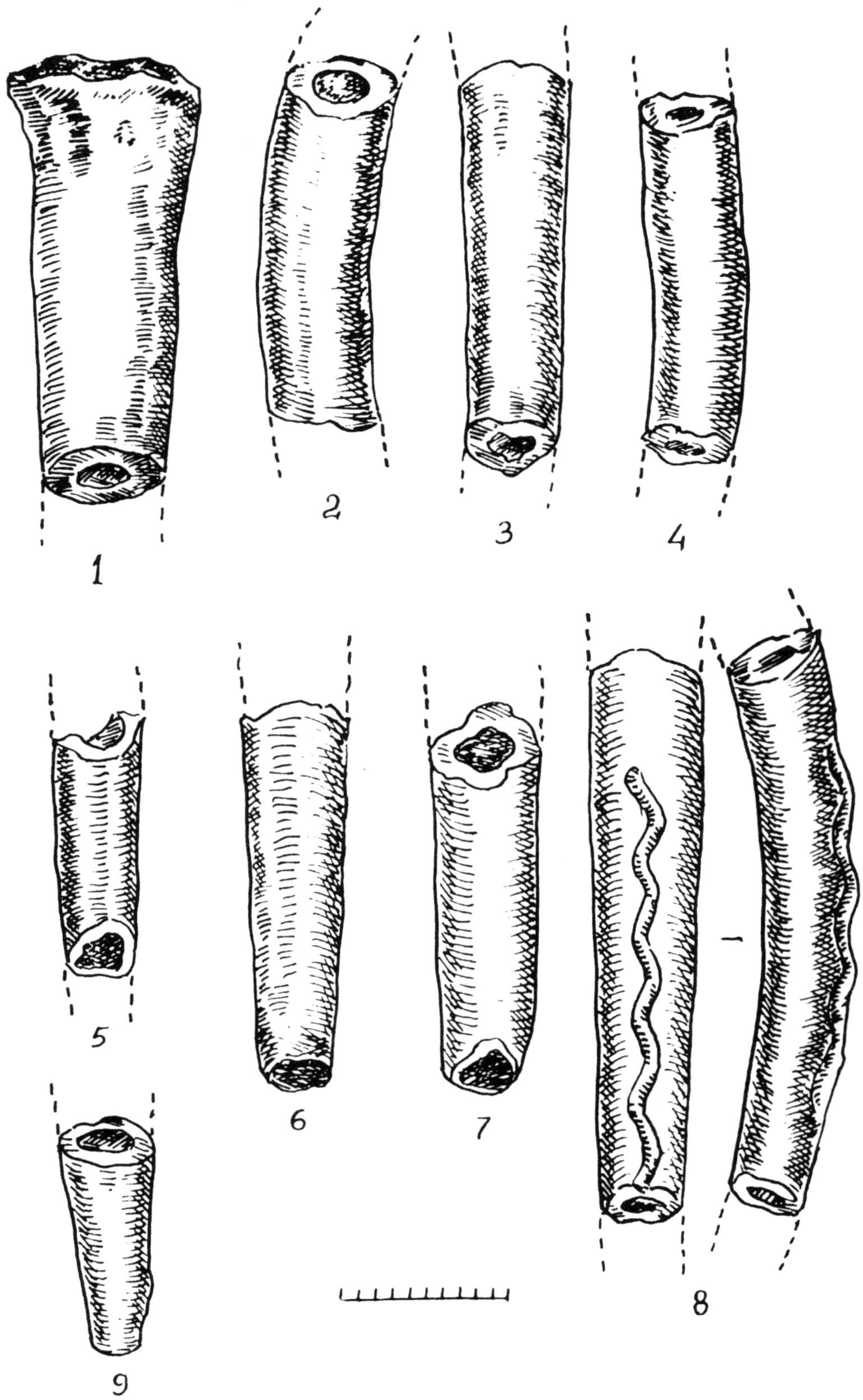

Fig. 51. "Mshvidobauri IV." 1-9 – fragments of clay nozzles for bellows;
8 – fragment of a nozzle with a snake design fashioned in relief.

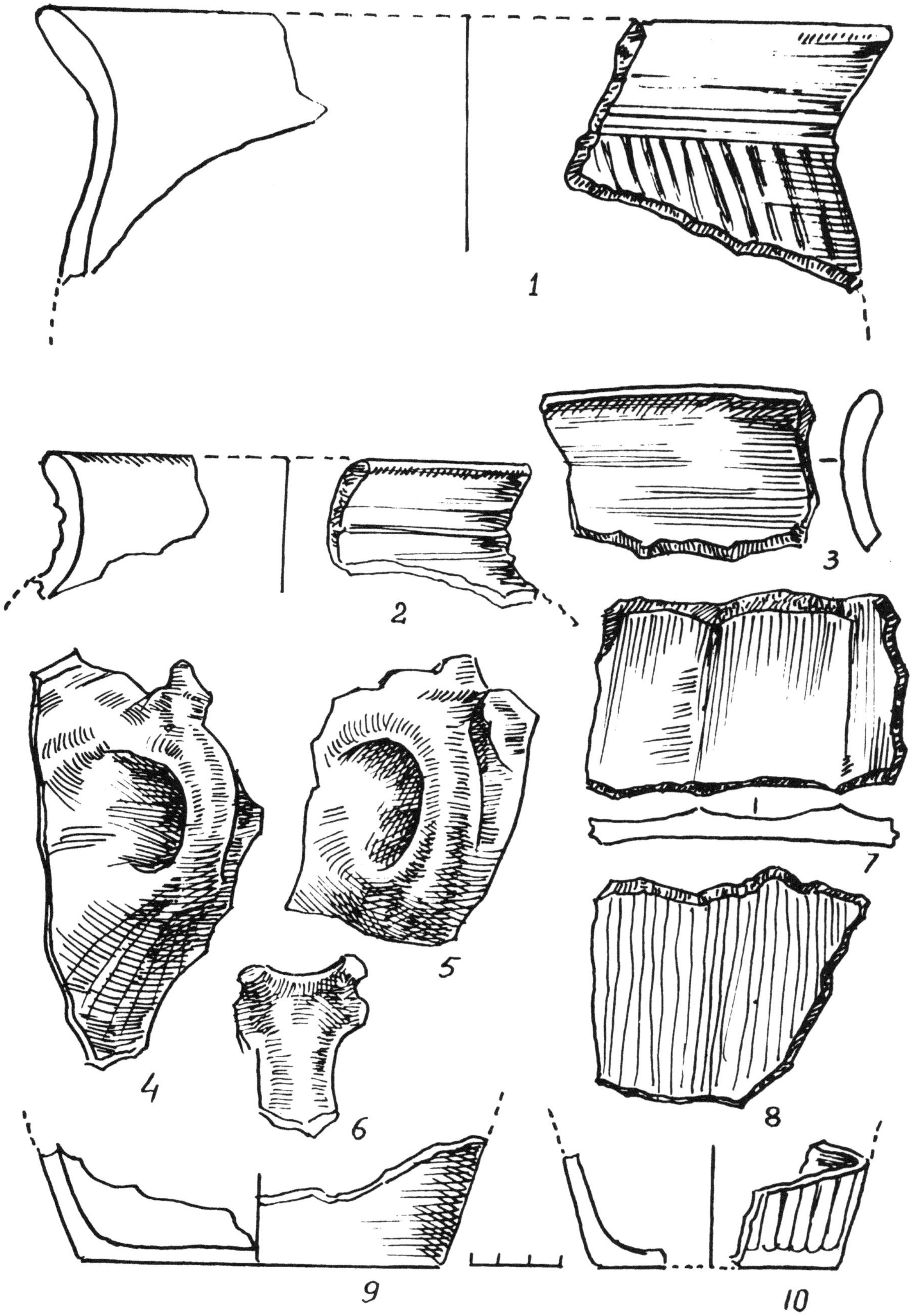

Fig. 52. "Mshvidobauri IV." 1-3 – fragments of vessels; 4-6 – fragments of vessels with zoomorphic handles; 7 – fragment of a thick-walled vessel with broad fluting; 9 – fragment of a vessel bottom; 10 – fragment of a thick-walled burnished clay vessel with broad fluting.

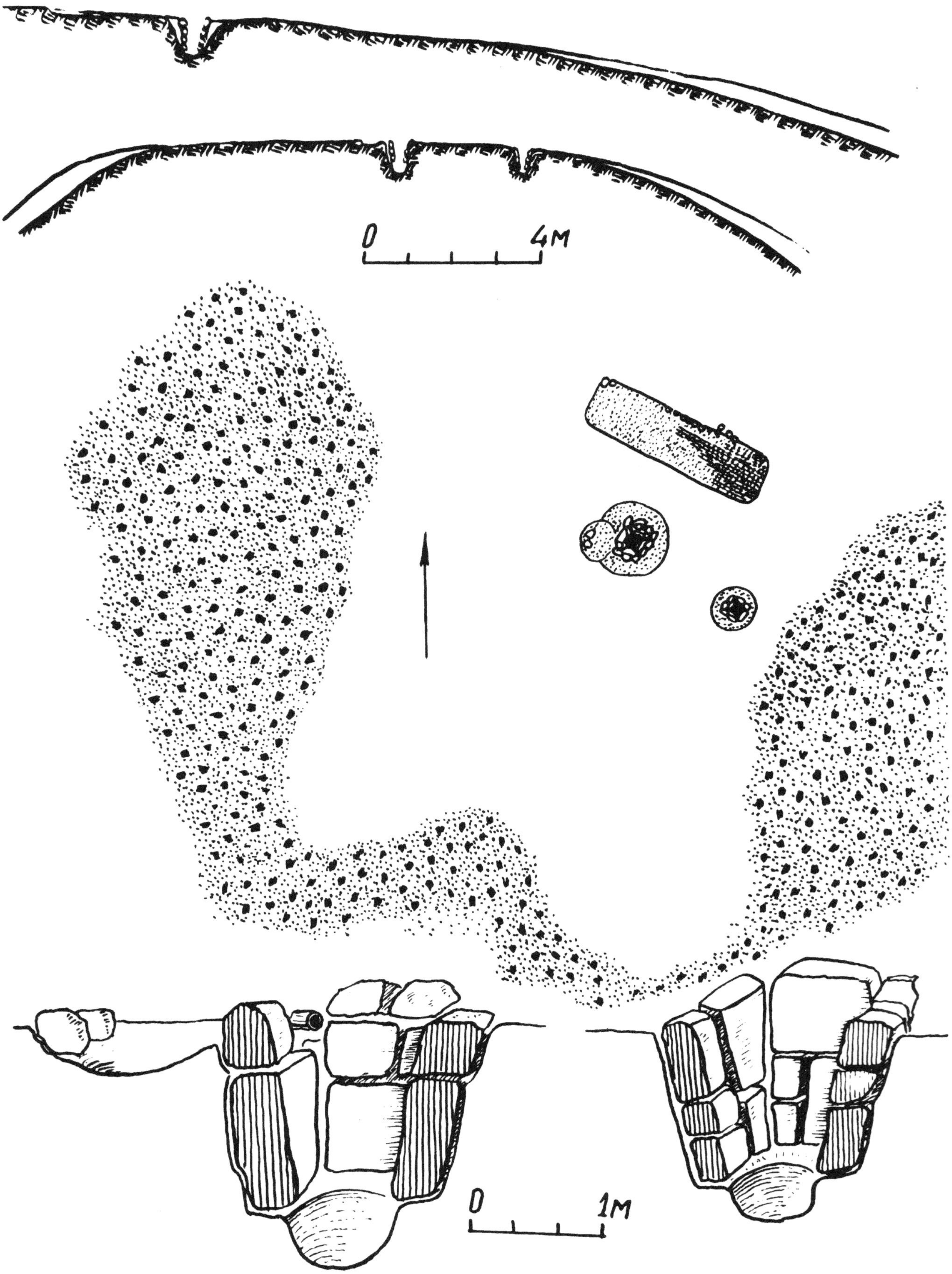

Fig. 53. Iron-smelting workshop "Nagomari I." Ground plan and sections.

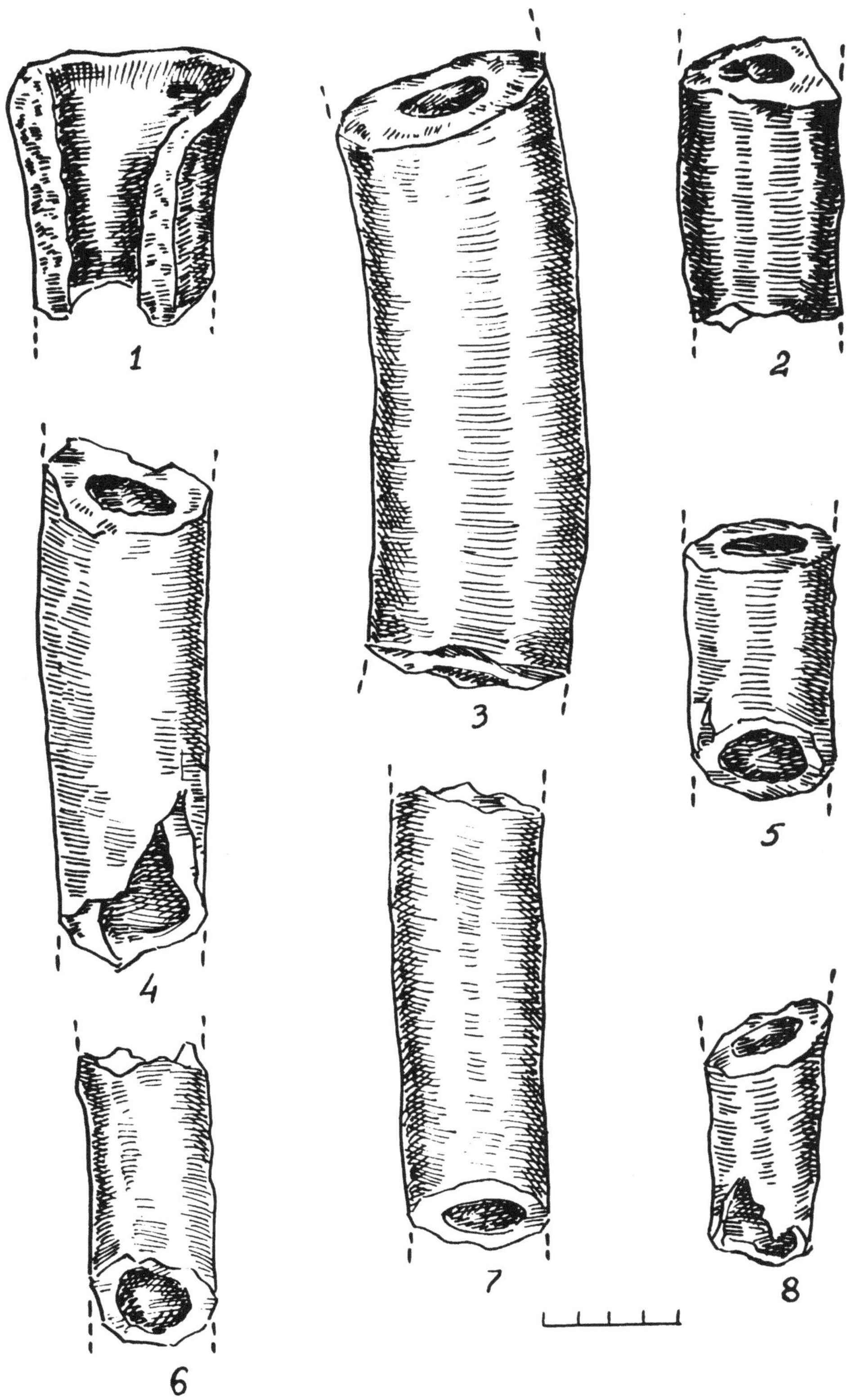

Fig. 54. "Nagomari I." Fragments of nozzles for bellows.

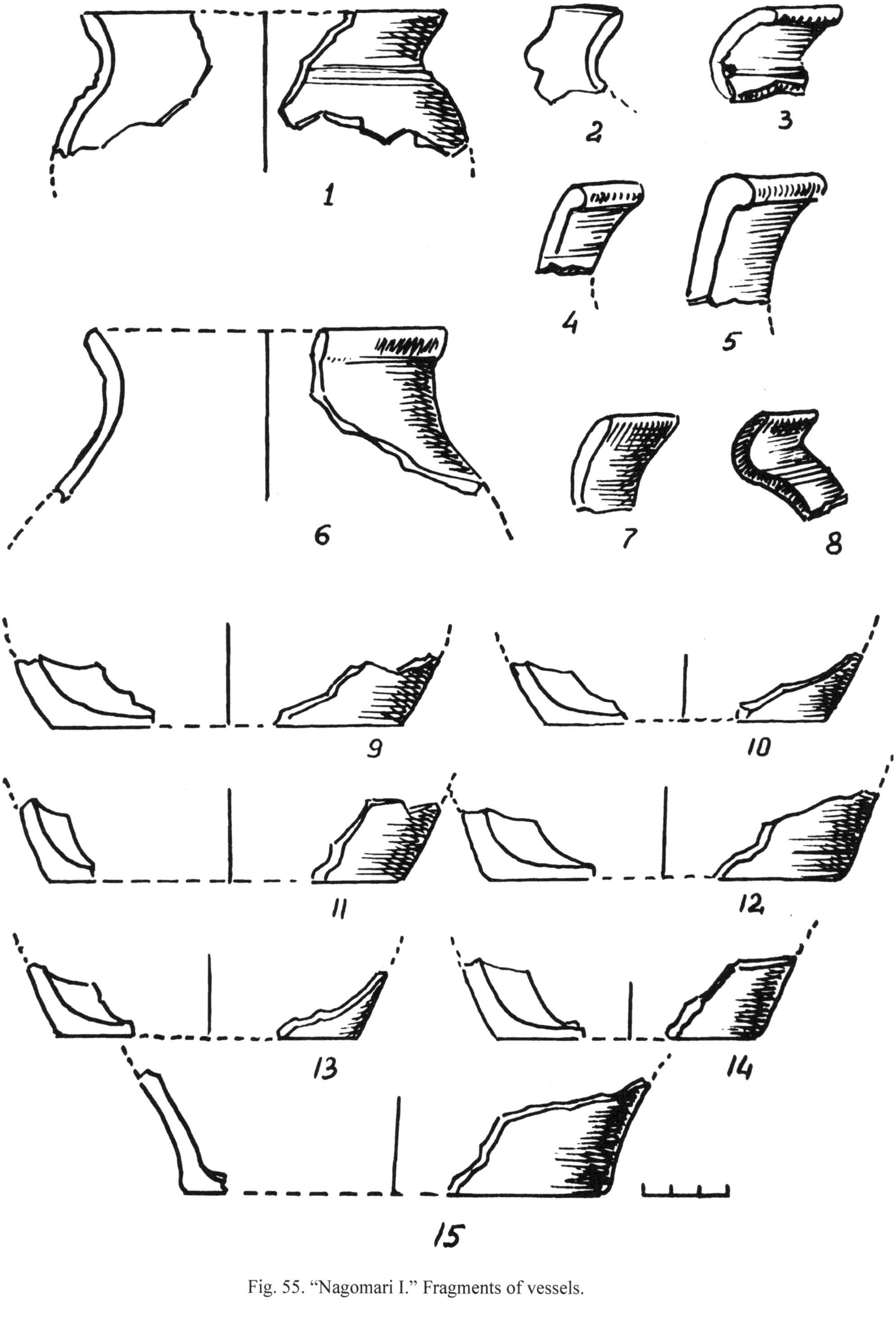

Fig. 55. "Nagomari I." Fragments of vessels.

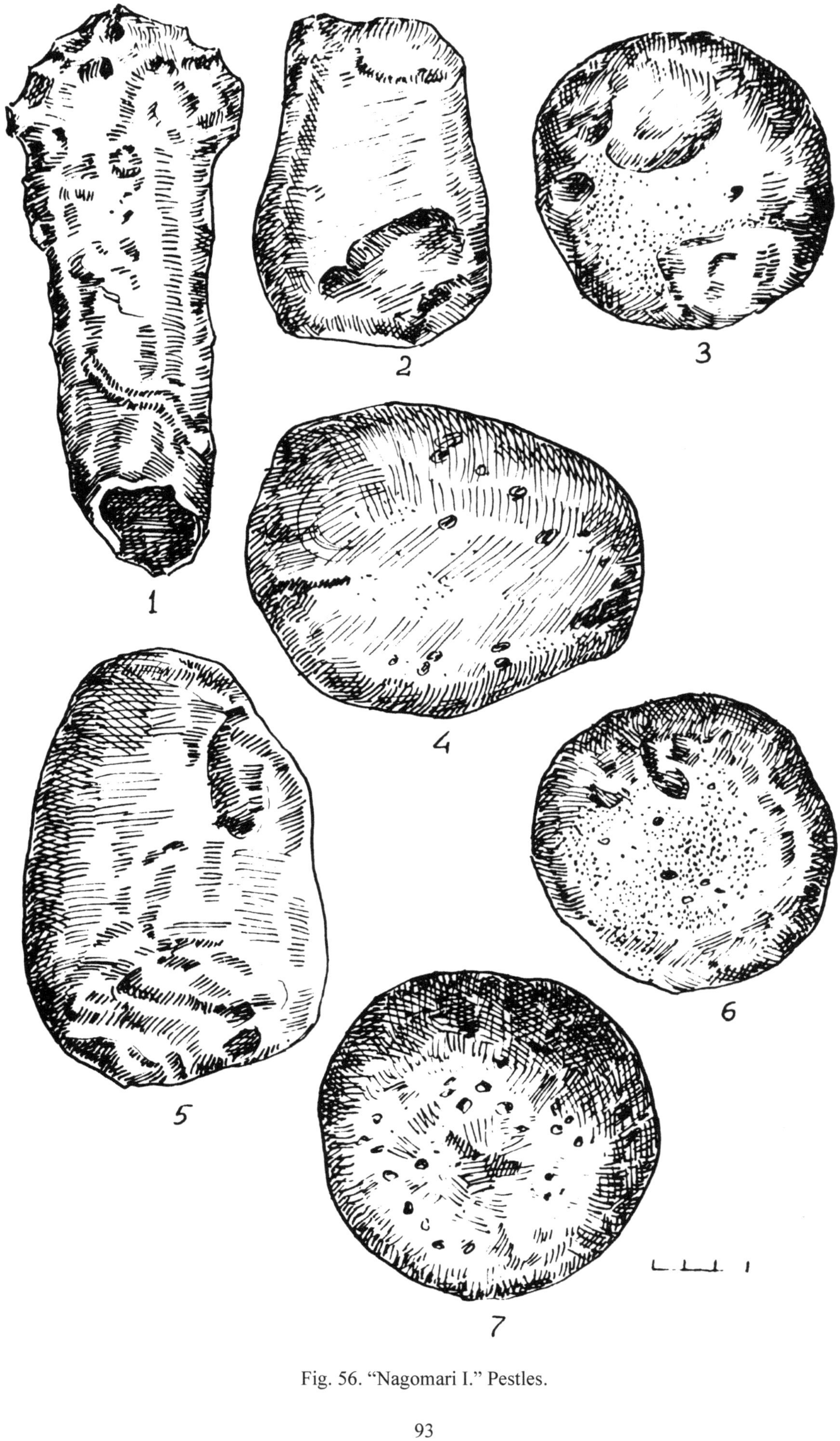

Fig. 56. "Nagomari I." Pestles.

shall not be far from the truth if we date this workshop to the tenth-ninth centuries B.C.

16. IRON-SMELTING WORKSHOP "NAGOMARI I, 2" (excavated in 1984)

The iron-smelting workshop "Nagomari I, 2" consisted of an iron-smelting furnace, a composite table (anvil) and a waste dump on the east slope of the site.

The iron-smelting furnace "Nagomari I, 1" is 230 cm east of furnace "Nagomari I, 1" (Fig. 53). It is arranged in a pit shaped like an upturned truncated pyramid. The bottom of the furnace is a hemispherical hollow plastered with a thick layer of fire-clay. The upper part of the pit is faced with rough-hewn stone and cobbles. A characteristic feature of this site is that during our 25 years of work devoted to excavating sites of ancient iron manufacture, we succeeded for the first time in registering the remains of the superstructure of an iron-smelting furnace: in the facing of the north wall we cleared two stones that were part of the base (beginning) of the upper vaulted part of the furnace.

In scientific publications it was suggested that the superstructures of ancient Colchian iron-smelting furnaces were built of stone and clay mortar. This supposition wasbased on the fact that when the pit of the iron-smelting furnace was cleared and the facing of the walls was found intact, a considerable number of scorched stones from the structure of the furnace were unearthed.[36]

The height of the surviving part of the furnace is 100 cm, the width at the top 57-77 cm, in the middle 35-40 cm, at the bottom 26 cm.

The inner space of the furnace was filled with clay, scorched stones from the structure of the furnace, lumps of slag and fragments of clay plaster, ash and crushed charcoal, as well as fragments of clay nozzles for bellows.

Of the composite table (anvil) there remained a scorched oblong plot partly covered by the composite table of the "Nagomari I, 1" workshop.

Part of the waste dump on the east slope and conjecturally in the south part of the upper levelled area terminated in front of the iron-smelting furnace: the dump was swept away by a bulldozer that was preparing the plantation road construction, and on the east slope it was badly damaged when tea shrubs were planted. Clearing the rest of the dump yielded lumps of slag with a greyish thick patina and pieces of ash-grey clay plaster, scorched stones from the structure of the furnace and the composite table, fragments of nozzles for bellows (Fig. 54); also non-diagnostic sherds of household ware.

[36] Khakhutaishvili. *Materials on the History ...* p. 16

Judging by its structure, characteristic features of the slag, as well as of the clay plaster, the "Nagomari I, 2" workshop belongs to the group of pre-Antique sites.

Analysis of charcoal from three levels of the iron smelting furnace yielded the following data:

1. Sample TB-438 taken from the bottom of the furnace from the level of the beginning of the wall lining indicated an absolute date 3410±50 = 1460 B.C. which corresponds to 3670±103 = 1720 B.C.
2. Sample TB-439, taken from the bottom of the furnace indicated an absolute date 3450±50 = 1500 B.C.;[37] according to Daymon *et al*. it corresponds to 3270±103 = 1770 B.C.[38]
3. Sample TB-440 being badly contaminated with clay and insufficient in quality would not lend itself to analysis.

It is does highly significant that both samples from the above iron-smelting furnace proved to be very close together, almost similar in age (according to the scale of semi-decay of C^{14}, 5730 years, the difference between the two samples is 40 years. i.e. 1460 B.C. - 1500 B.C., while according to the table of Daymon *et al.* it is 50 years: 1720 B.C. - 1770 B.C. At the same time, the time when this workshop was in operation, a very early period, rare even for Colchis, is of great interest to modern archaeological science. According to this index, it is second only to the "Mshvidobauri II" iron-smelting furnaces (TB-432). 3483±125 - 1533 B.C. According to Daymon *et al.*, - 3760±103 = 1810 B.C.), so far the oldest site, and the oldest known workshop.

Written sources, as well as archaeological facts such as finds of iron objects in strata of the first half of the second millennium B.C.,[39] with, at the same time, the almost complete absence of metallurgical sites in regions where iron articles were plentiful make us think that ancient Colchis, (the Eastern and South-Eastern Black Sea littoral) occupied a particular place in the emergence and development of early iron metallurgy (a concrete discussion of these problems will be presented in the conclusion to this book).

[37] Data from the Dating Laboratory, Tbilisi University. (Burchuladze, and Togoidze).

[38] Arslanov. *op. cit.*, p. 1161, pl.2.

[39] J.C. Waldbaum. "The First Archaeological Appearance of Iron and the Transition to the Iron Age." *The Coming of The Iron Age*. New Haven, 1980, pp. 69 ff. J. Siegelova. "Gewinnung und Verarbeiten von Eisen in Hethitischen Reich in Jahrtausend v.u.z." *Annals of the Naprstek Museum* 12, Prague, 1984, pp. 71-168.

CHAPTER IV

THE KHOBI-OCHKHOMURI MANUFACTURING CENTRE

The Khobi-Ochkhomuri manufacturing centre in the Colchian area of ancient iron manufacture was discovered in 1960 in the Chkhorotsku district, Georgian SSR, by the joint archaeological expedition of the I.A. Djavakhishvili Institute of History, Archaeology and Ethnology and of the N.A. Berdzenishvili Research Institute of the Academy of Sciences of the Georgian SSR in Batumi.[1] The archaeological survey of the site was carried out in 1972 by a special group[2] of the Colchian archaeological expedition which excavated three sites and published a preliminary report on the work done by its members.[3]

The manufacturing centre was located in the gorge of the Ochkhomuri River, a left-bank tributary of the river Khobi which (Fig. 56A) flows into the Black Sea at Kulevi, on the territories of the villages Pirveli Choga, Meoré Choga, Nakiani, Napichkhou, Ledersalé, Khabumé, and others. The distance from the sea shore to the iron-manufacturing centre is about 120 km.: this circumstance raised the problem of the location of the source of raw materials.

The basin of the Ochkhamuri consists of the following brooks and streams: Dichké, Shubé, Choga gali, Kirdgimi, Gadidi (Gal-didi), Ogvabé, Okarkalé, Lepataré, etc.

In this area the soil is the same as in the entire subtropical foothill zone of the Eastern Black Sea littoral, with the only difference that here, beneath a stratum of yellow diluvial clay, we observe in places limestone and marl deposits into which the Ochkhamuri river, as well as some of its tributaries, have cut their beds.

The terrain here is broken, hilly, and covered with profuse vegetation. According to old residents it was all wild forest-land in the past, with almost all the foliage trees typical of the subtropical zone.

This ancient metallurgy group consists of over 50 sites. They are identified by their waste dumps, part of whose surface is, for several reasons, free of vegetation.

The village of Pirveli Choga lies on both banks of the Ochkhamuri, but sites of ancient iron manufacture have not been identified on the banks.

The same situation has been observed in other Colchian metallurgical centres: remains of iron-smelting workshops occur only on the banks of streams and gullies, but not on the banks of the main rivers in the hydro-system.

In the village of Pirveli Choga, ancient metallurgy sites occur on both banks of the river: over 15 are marked on the archaeological map. Three of them have been excavated by us.

A survey of the territory of the production centre revealed that in the middle of Pirveli Choga village there had been ancient settlements dating back to the Early Iron Age. This was the first discovery of a settlement surrounded by ancient iron production sites. Unfortunately, a number of practical reasons prevented us from excavating this site. Future fieldwork will show whether, or how far, there was any interdependence between these settlements and the iron-smelting workshops. This supposition is not based on geographical considerations alone. The fact that when settlement "Choga I" was investigated, lumps of iron slag were found among other materials in a road cutting into a cultural layer 1.5 metre thick near the bridge across the Ochkhomuri river. This also occurred in another settlement, almost adjacent to the former, known in scientific publications as settlement "Choga II," or settlement "Choga-Kakutskha."[4]

1. IRON-SMELTING WORKSHOP "CHOGA I" (excavated in 1972)

The iron-smelting workshop "Choga I" was identified on the land ploughed by team No.5 of the Choga village collective farm, on a tea plantation in a locality called Shkaga, to the right of Lepatare stream, near a road below the former holding of a certain Lomia, on a south-facing slope.

Unfortunately, a few months before the arrival of archaeologists the site was badly damaged when the plantation road was being built: the bulldozer swept away the most important part the plot where the iron-smelting furnace had been, and part of the waste dump. (Fig. 57).[5]

The only thing that remained of the "Chaga I" iron-smelting furnace is a burnt red stain on its north wall, and scorched stones from the structure of the furnace scattered here and there.

The composite table for working the spongy mass was identified to the north of the conjectural site of the

[1] The site was discovered thanks to the information received by I.A. Gzelishvili from G.G. Shengelia, director of studies at the secondary school in the village of Pirveli Choga.

[2] The personnel of the group: D.A. Khakhutaishvili (head), N.V. Khazaradze, N.D. Khakhutaishvili, A.M. Djavelidze, G.G. Shengelia, T.M. Machavariani, E.N. Antidze.

[3] Khakhutaishvili. *Materials on the History …*, Tbilisi, 1980, pp. 3-38.

[4] Khakhutaishvili. *op. cit.*, pp. 16,29.

[5] Gzelishvili, *Iron Metallurgy in Ancient Georgia*, p. 58, Fig. 17 A; Figs. 17, 20, 22, 23, 28.

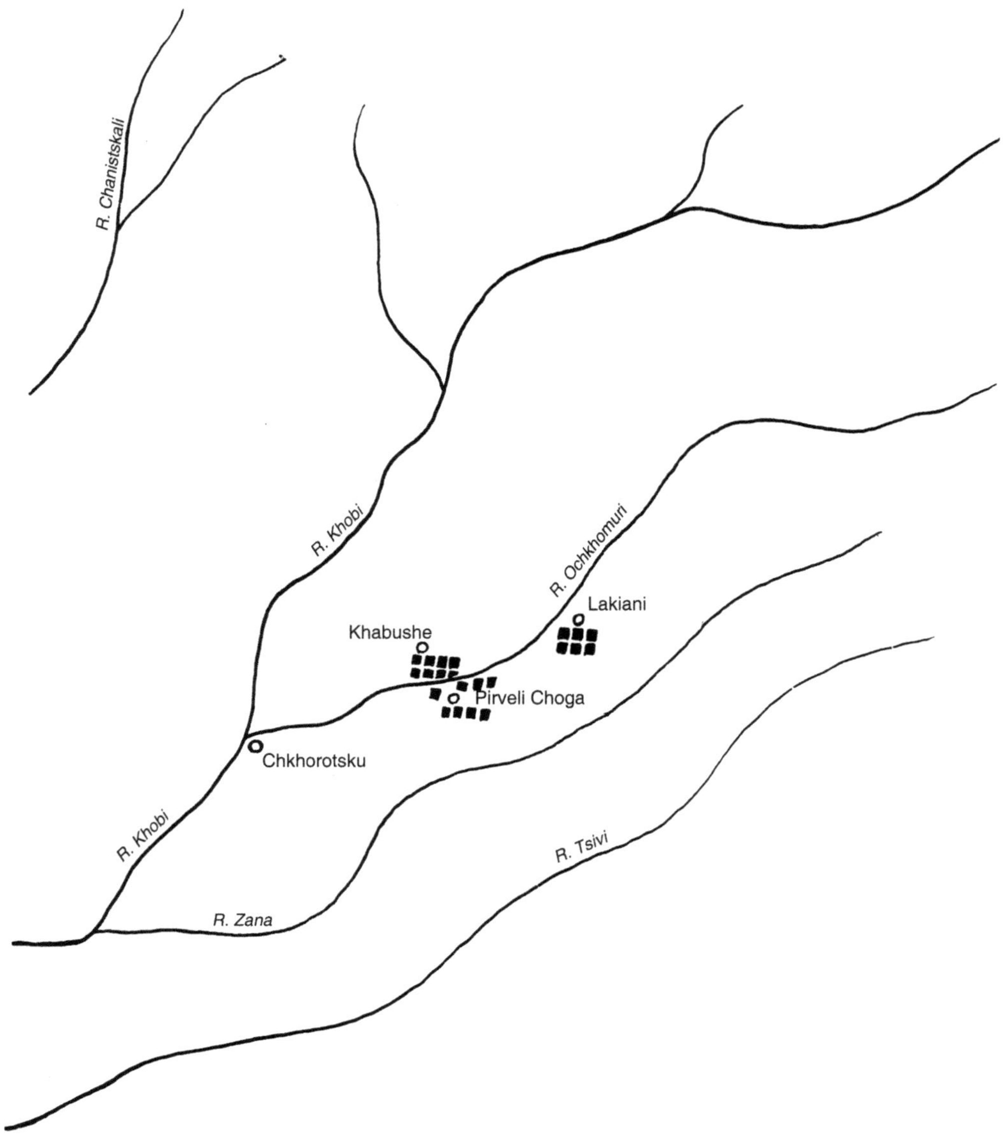

Fig. 56A. Location of separate groups of iron-smelting workshops of the Khobi-Ochkhomuri production centre.

furnace, 100 cm from it, but part of the table was destroyed when the road was laid (Fig. 57). The table was 250 cm long and 80 cm wide; it was bordered by a row of small stones, part of the inner space was paved with flat cobbles, another part with fragments of clay plaster (Fig. 57). The table was damaged when the tea plantation was laid out here.

Along the north and west sides of the table there was a trench burnt red 260 cm long and 35 cm wide (Fig. 57). The trench was filled with crushed charcoal ash and baked clay in which were separate pieces of haematite.

All these details prove that we had succeeded in identifying a new element in a Colchian iron-smelting workshop, namely a platform where ore was given a preliminary roasting. Such elements are not characteristic of iron-producing centres situated at a distance of 10-30 km. from the sea shore, where magnetitic sands were used as raw material.

The waste dump contained a great many lumps of slag (about 5 m^3) with a thick greyish patina, scorched stones from the structure of the furnace, the composite table, and from the platform for preliminary roasting of the ore; also ash and pieces of charcoal, fired clay, ash-grey fragments of nozzles for bellows and of pottery (Fig. 58).

Thus, excavation of the remains of the "Choga I" iron-smelting workshop has revealed a new element a platform on which the ore was preliminarily roasted.[6] Such platforms were most probably the prototype of later ore-roasting furnaces found in many parts of medieval Georgia.[7] We have at our disposal some data to help us determine the date at which when "Choga I" functioned:

[6] Khakhutaishvili, *op. cit.*, pp. 16, 29.

[7] Gzelishvili, *Iron Mettalurgy in Ancient Georgia*, p. 58, Fig. 17a; p. 52. Figs. 17, 20, 22, 23, 28.

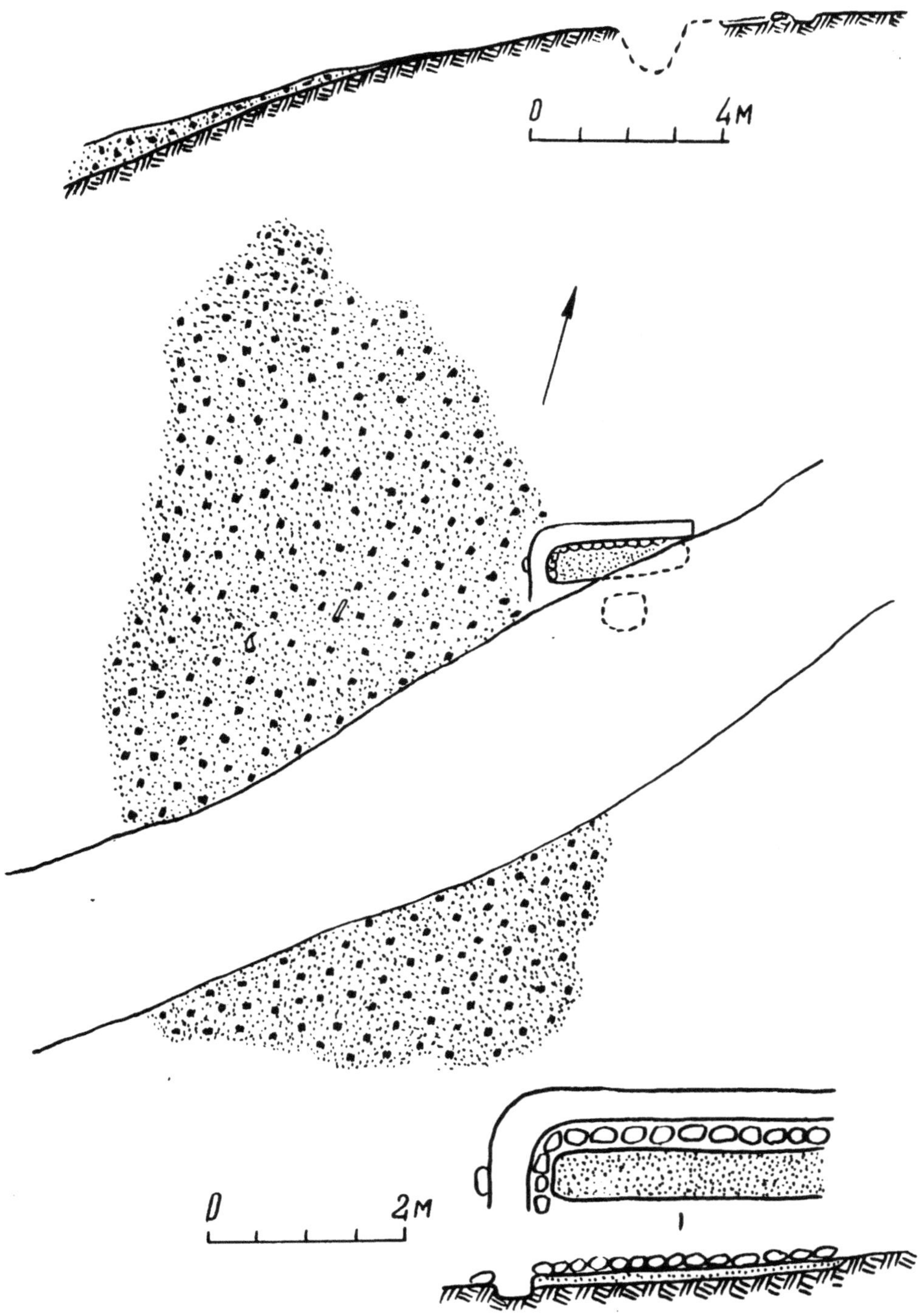

Fig. 57. Iron-smelting workshop "Choga I." Ground plan and sections.

a) the iron slag with a thick greyish patina shows that the site belongs to the pre-Antique period;

b) the earthenware vessel fragments that were found during the excavation of the surviving remains of the iron-manufacturing workshop are similar to ceramics of the first quarter of the first millennium B.C. Unfortunately, no other indications of the date of this workshop have survived.

2. IRON-SMELTING WORKSHOP "CHOGA II" (excavated in 1974)

The remains of the "Choga II"[8] iron-smelting workshop were identified on the tea plantation belonging to the village of Pirveli Choga, not far from a shed built near a spring of water, at the confluence of two brooks and gullies, the Lepataré and the Lestkvitaria, on the right bank of the latter. The slope where the waste dump is situated has a south-western exposure. The centre of the

[8] For a preliminary publication of the site, see: Khakhutaishvili, *op. cit.*, pp. 10-19, Figs. 1-6.

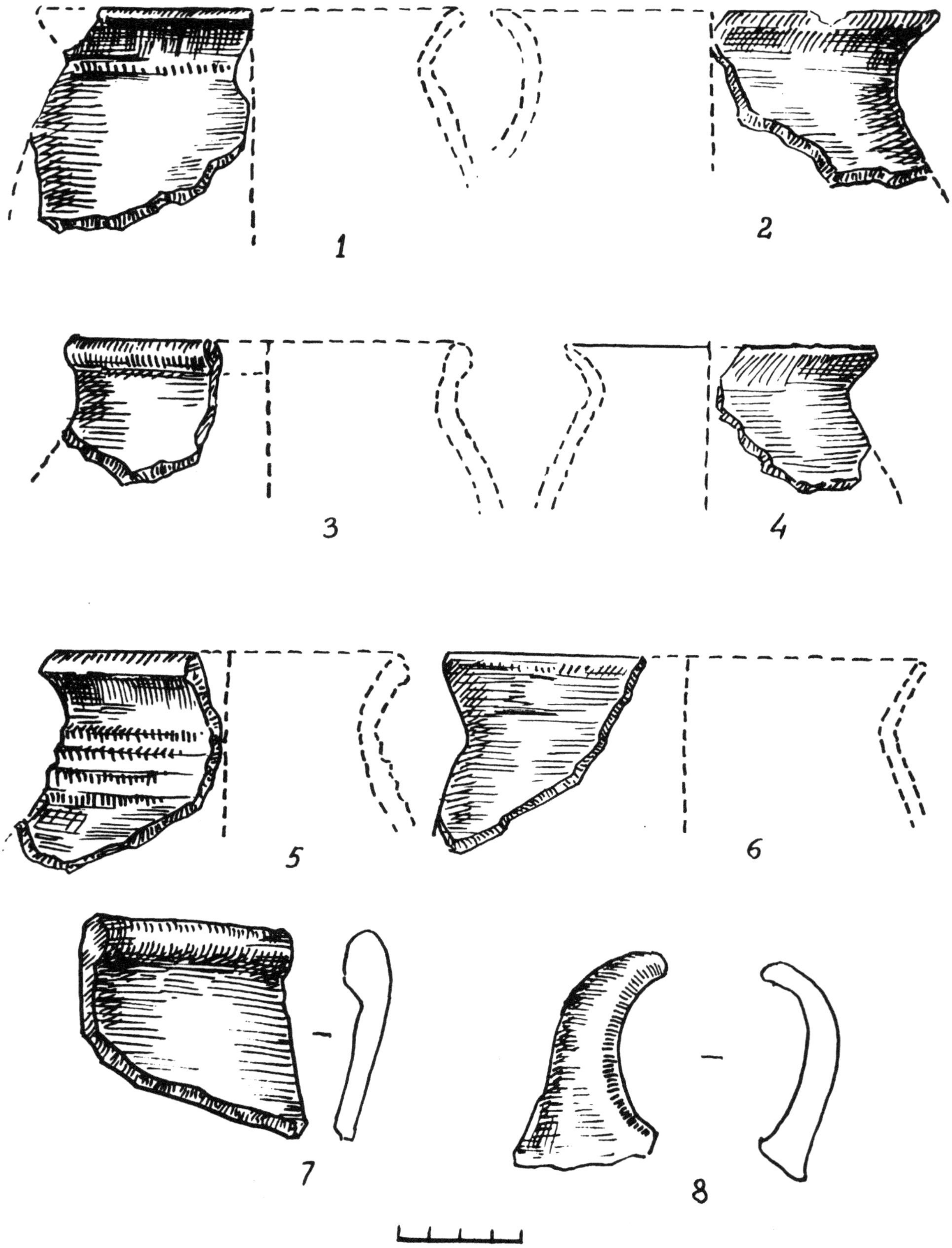

Fig. 58. "Choga I." Fragments of vessels.

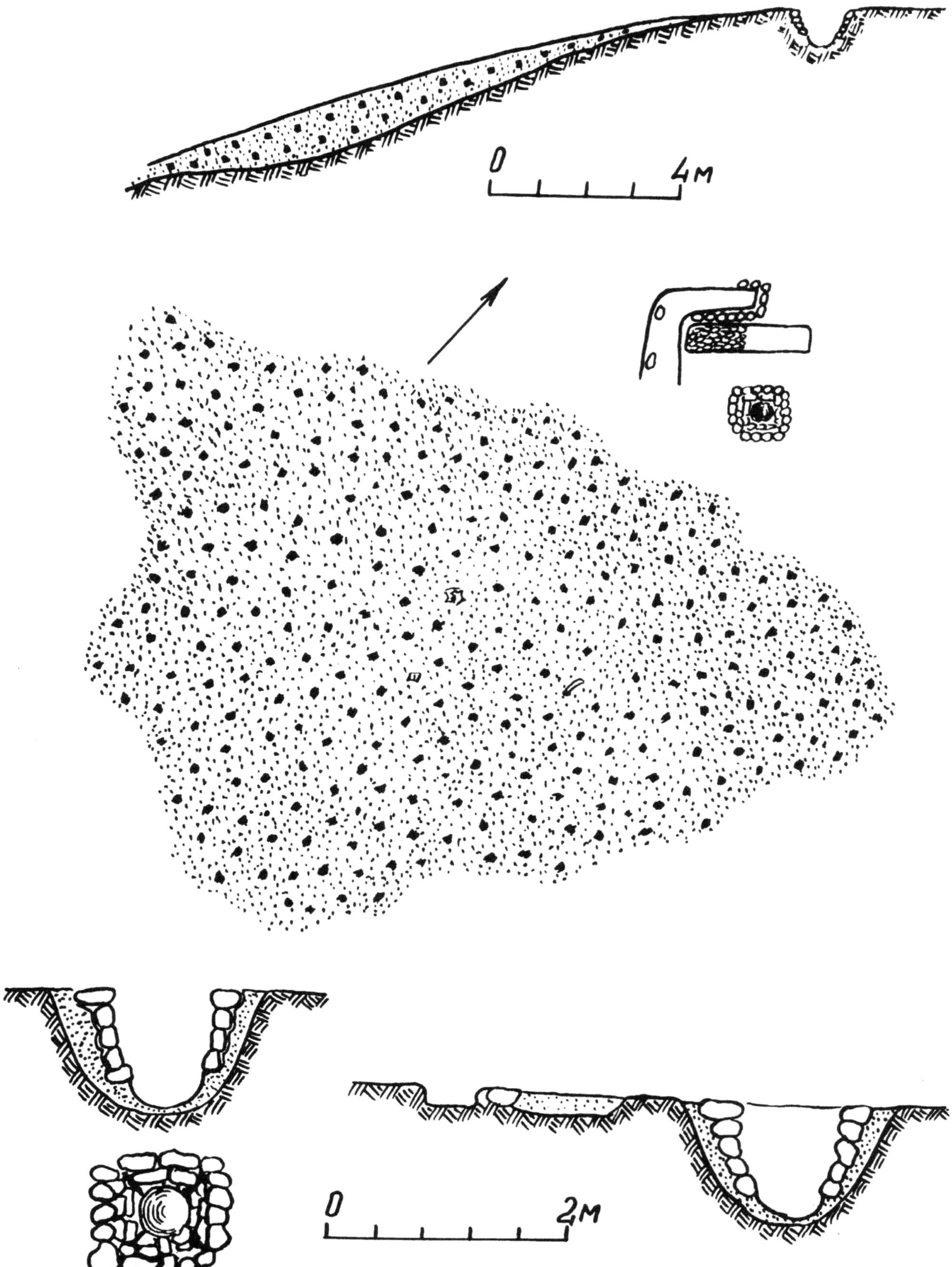

Fig. 59. Iron-smelting workshop "Choga II." Ground plan and sections.

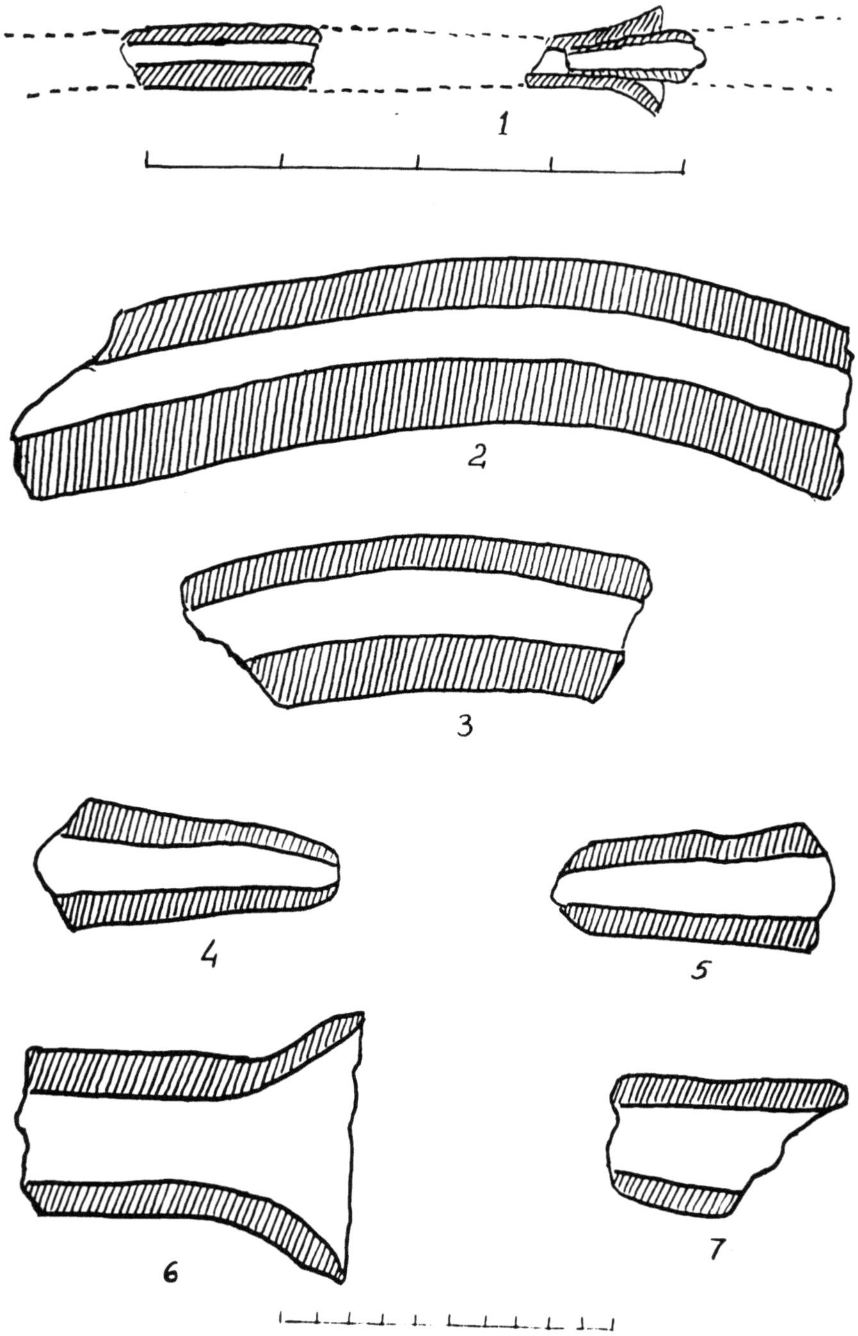

Fig. 60. Sections of nozzles for bellows.

dump consists mostly of lumps of iron slag and has no vegetation on it. According to a local resident, several attempts to plant tea bushes there proved fruitless. The main part of the territory, as a preliminary survey showed, revealed traces of a workshop covered with thick rows of tea bushes.

The territory was cleared of tea bushes, and remains of the workshop were laid bare. When the humus and a layer of yellow diluvial clay were removed, the principal parts characteristic of North Colchian iron-smelting workshops were exposed to view at a depth of 30-40 cm (Fig. 59).

The "Choga II" iron-smelting furnace (Fig. 59), 5-6 metres from its waste dump and situated on a plot higher uphill, was a pit shaped like an upturned truncated pyramid, faced from top to bottom with rough-hewn stone and cobbles.[9] At a depth of 15-20 cm, the ground around the furnace was burnt red by the intense heat.

Within, the iron-smelting furnace was filled with clay, scorched stones from the structure, lumps of slag and pieces of charcoal, fragments of clay nozzles for bellows, ash and other admixtures. At the bottom of the furnace was a layer of crushed charcoal mixed with clay.

[9] The furnace is preserved *in situ*.

The bellows were installed on the north-western side of the furnace, which is attested by a small, round, red-burnt space at the edge of the furnace.

The height of the surviving part of the furnace is 105 cm, the width at the top 70-90 cm, at the bottom 20 cm, in the middle 57-60 cm The facing of the furnace has remained intact; despite this, 0.15 cubic metres of stones were extracted from the interior of the furnace, evidently from the ground above part of it. This suggests that the overall height of the furnace may have been 130-140 cm, or a little higher.

The composite table (anvil) situated north-west of the furnace was built partly of stones and pieces of slag (Fig. 59). The table, oriented from south-west to north-east, was 260 cm long and 60 cm wide. The area of the paved north-western part of the table was 100×60 cm All the area of the table bears traces of having been subjected to high temperatures; there is a noticeable layer of fine ferrous powder, evidently formed when the spongy mass brought from the nearby iron-smelting furnace was being forged. The ground beneath the table is burnt red.

To the north-west, the composite table is bordered by a trench, 39 cm wide at the beginning, and 80 cm at the end. The wide part of the trench reaches the furnace. The depth of the trench is 15-17 cm Here and there along its edge a border of cobbles has survived in part (Fig. 59). The trench was filled with crushed charcoal mixed with clay and separate fragments of clay nozzles for bellows. A few lumps of haematite were also found. At the bottom of the trench, close to the composite table, there was a small pit, the purpose of which is not quite clear.

The trench was, most probably, intended for preliminary roasting of the ore (haematite).

The waste dump was on a slope facing south. Clearing the dump yielded numerous lumps of iron slag with a thick greyish patina, as well as pieces of clay plaster of the furnace, ash-grey in colour; fragments of nozzles of clay bellows (Fig. 60), and sherds of burnished pottery (Fig. 61); also zoomorphic handles, flat-bottomed pots and vases with a rounded rim. The earthenware vessels are, as a rule, fashioned on a potter's wheel from well-washed and well-refined clay with an admixture of sand. Among the nozzles for bellows found in the dump, some are noteworthy: they have funnel-like openings meant to ensure airtight fixing to the bellows; some of the nozzles are curved.

The dump also contained a much-hollowed mortar made from a flat stone. Such mortars are usually considered to have been used for pounding flux and ore. They occur in the materials of nearly all the iron manufacturing centres of ancient Colchian metallurgy, as well as of metal-smelting workshops of a later period.[10] The dump also yielded pieces of coal and lumps of ore (haematite), ash and scorched stones.

Thus, the "Choga II" iron-smelting workshop had all the component parts characteristic of the given group of North Colchian workshops: the iron-smelting furnace, the composite table (anvil), a place for the preliminary roasting of the ore, and waste dump.

The dating of this site was facilitated by the ceramics, by the results of radiocarbon and archaeomagnetic analyses.

The ceramic material from "Choga II" is almost identical with some of the material from the ancient settlement "Kulevi IV," at the mouth of the Khobi river,[11] "Namcheduri III" near Kobuleti, "Tsikhisdziri IV," the Vani site (the pre-antique stratum), the upper and middle strata of the "Nosiri" settlement, etc. A careful comparison of materials from various sites of maritime Colchis gave us grounds to assign "Choga II" to the 10th-9th centuries B.C.[12] According to archaeomagnetic data obtained by Z.A. Chelidze, "Choga II" functioned before "Choga I" and after "Askana IV" (see above, relevant parts of Chapters II and III).[13]

According to the results of the radiocarbon analysis (done at the Dating Laboratory of Tbilisi University by Burchuladze and Togonidze) of a piece of charcoal taken from the trench for roasting the ore, "Choga II" functioned about 646 B.C. (2596±45) which corresponds to 710 B.C. (2660±102) in the table compiled by Daymon *at al.* Since other data, especially purely archaeological ones, assign "Choge II" to an earlier period, we suppose that the samples of charcoal taken for analysis from "Choga II" were somewhat contaminated (which would account for the late date of the sample). We therefore provisionally assign the workshop "Choga II" to the l0th-9th centuries B.C. We are incidentally of the opinion that workshops "Choga I" and "Choga III" can be assigned to the same period.

3. IRON-SMELTING WORKSHOP "CHOGA III"

The remains of the iron-smelting workshop "Choga III"[14] were identified on an oblong hillock between two gullies, at the bottom of which flow the streams Lepatare gali and Letskhvitaria gali; it lies west of "Choga II" iron-smelting workshop, on part of the tea plantation Khantska, near shed No. 6 belonging to a team of the

[10] B.E. Degen-Kovalevsky, "On the Materials from the Iron-Smelting Furnace in Chuber." In: *Svania on the History of Iron Manufacture in Transcaucasia*. Proceedings of 1935, 120, pp. 270-272, Figs. 15-17.

[11] Excavated by D.A. Khakhutaishvili, 1972-1973, as a member of the Colchian Archaeological Expedition (headed by Prof. T.K. Mikeladze).

[12] Khakhutaishvili, *op.cit*., pp. 30-33.

[13] Z.A. Chelidze, "Results of Archaeomagnetic Investigations of Some Archaeological Sites." *Problems of Ancient History*, X, pp. 142.

[14] For a preliminary publication of the site see: Khakhutaishvili. *op. cit.*, pp. 19-28, Figs. 7-8.

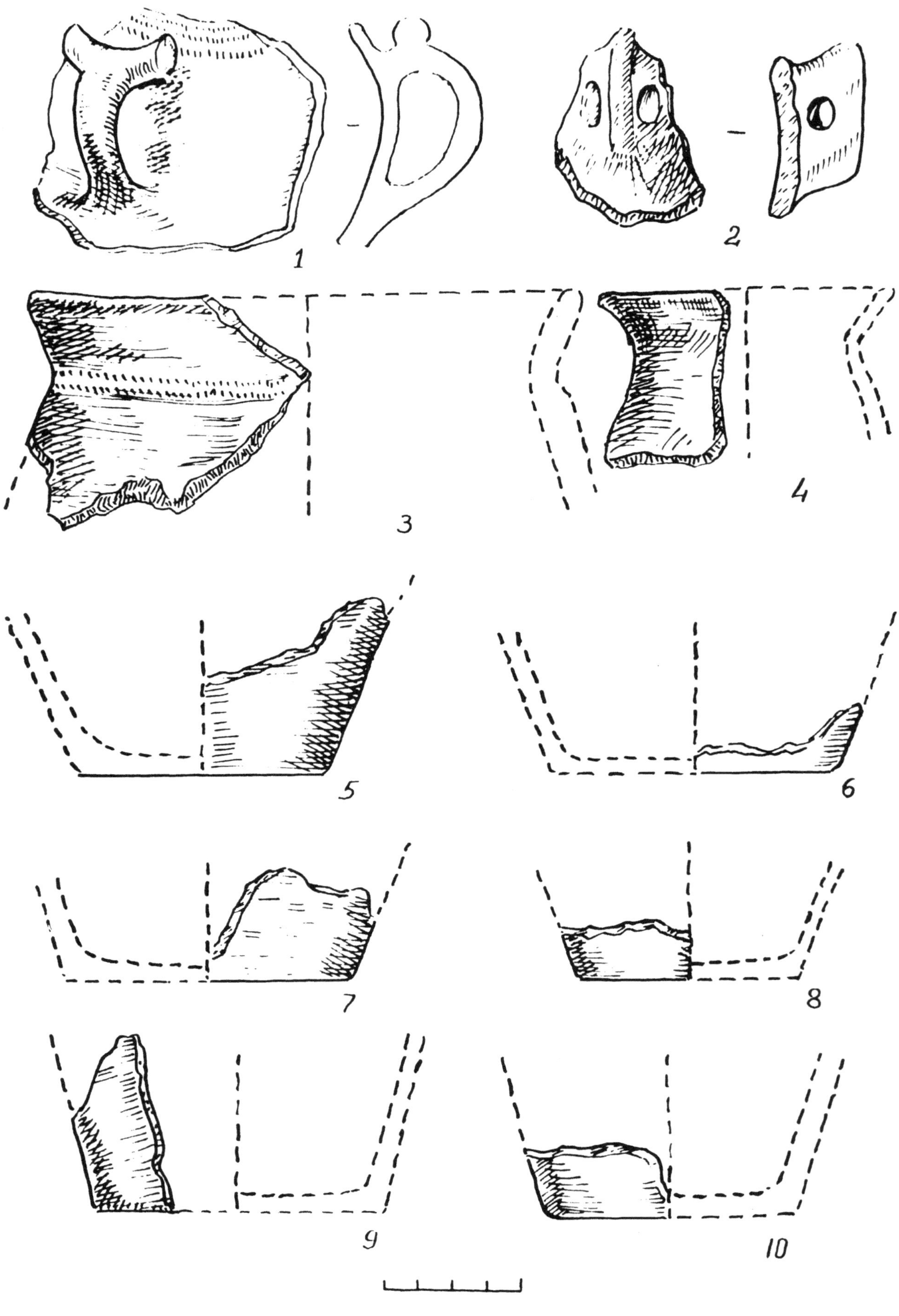

Fig. 61. "Choga II." 1 – fragment of a vessel with a zoomorphic handle; 2 – lug of a vessel; 3-10 – fragment of vessels.

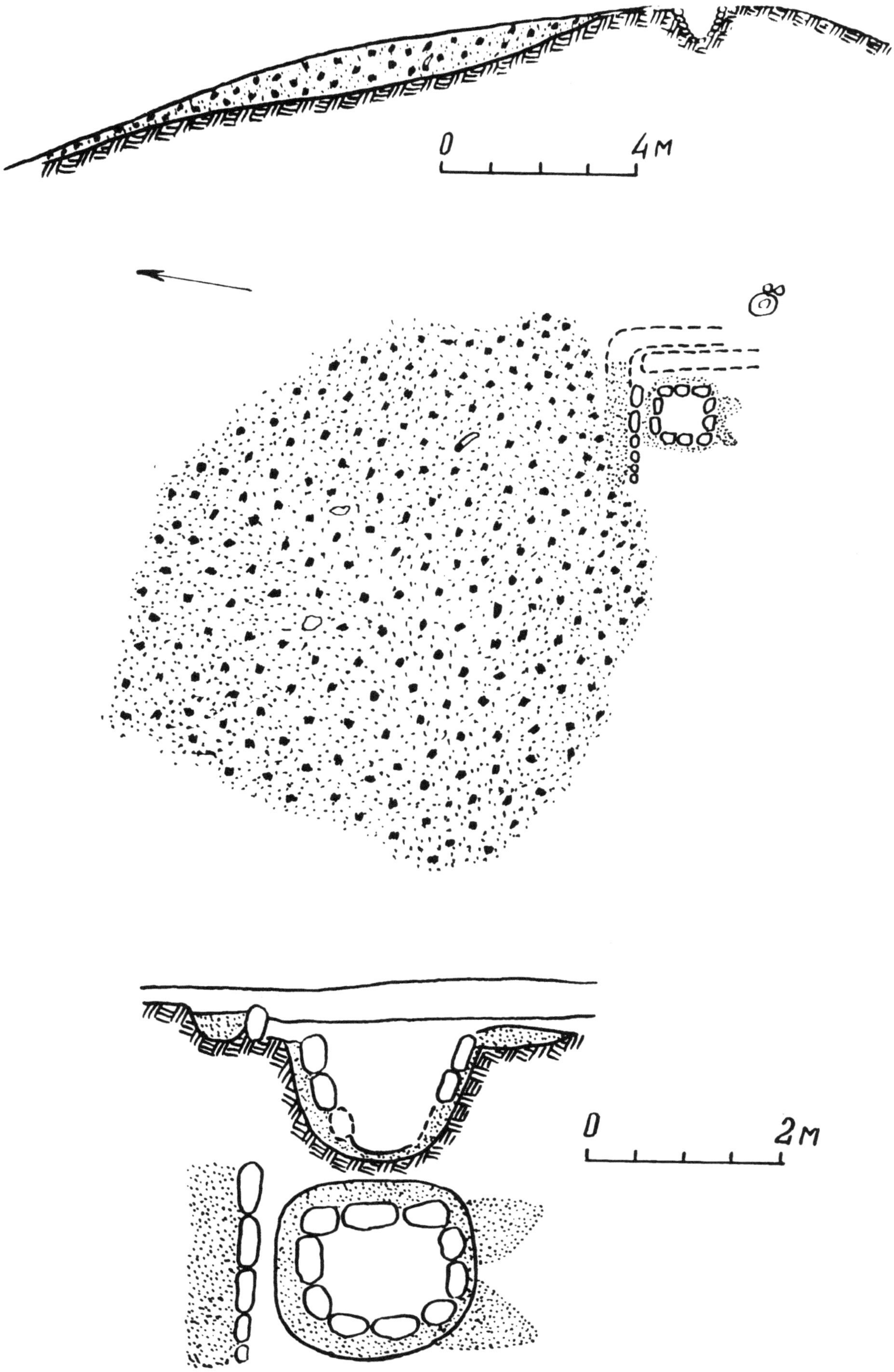

Fig. 62. Iron-smelting workshop “Choga III.” Ground plan and sections.

Pirveli Choga village collective farm, 500 m. to the south-east of the shed.

Near the remains of the workshop, to the east of it, there is a steep precipitous slope down which the greater part of the waste was thrown. The slope is overgrown with a thicket of fern, various climbing and creeping plants and trees. To the west, the area on which the remains of the iron-smelting workshop were discovered slopes gently towards the Lepatare gully. Curiously, though the collective farmers had succeeded in planting tea bushes there, the bushes would not thrive and turned yellow. It is just this aspect of the tea bushes that made us seek the cause of the above anomaly. In this area we found lumps of slag and pieces of clay plaster scattered among the bushes.

A few months before our arrival, the contours of a future plantation road had been marked out by a bulldozer on the very crest of the rise, across the conjectured location of the iron-smelting workshop (the furnace, the composite table, etc.). After the tea bushes had been cut down and removed from the area of the workshop and after the humus was also removed, the contours of the component elements of the workshop came into view (Fig. 62).

The iron-smelting furnace "Choga III," or more exactly the part below ground, was a pit shaped like an upturned truncated pyramid (Fig. 62). The remains of the furnace were overlain by a layer of yellow clay 30 cm thick. The walls of the furnace had been faced with stones of various shapes and sizes, but except for the two upper courses, had fallen in and filled the interior space of the furnaces where, besides the scorched stones from the structure of the furnace, there were lumps of slag and pieces of clay plaster, burnt clay and crushed charcoal. The soil around the furnace at a depth of 15-20 cm, had been burnt reddish by the intense heat.

The dimensions of the surviving part of the furnace are: height 80 cm, width at the ground level 65-70 cm, in the middle 40-50 cm, at the bottom 20-25 cm.

The bellows were fixed to the south-west corner of the furnace, which is attested by two round patches of burnt soil (Fig. 62). East of the furnace, 140 cm away from it, there was a hemispherical pit burnt red, 45-50 cm in diameter, and 20 cm deep. The purpose of this pit is unclear.

The composite table (anvil) was damaged when the road was being built; there remained only a rectangular area burnt red, on two sides of which are traces of a trench for preliminary open roasting of ore. There had also survived a single course of dry stone masonry bordering the trench for roasting ore (Fig. 62). At the bottom of the trench a thin layer of crushed charcoal was found, but there was not enough of it for a radiocarbon test.

The waste dump was on a gentle slope west of the furnace; it contained a large amount of ash, of lumps of iron slag with a thick greyish patina, ash-grey pieces of clay plaster of the furnace, potsherds and scorched stones from the structure of the furnace and of the composite table. The texture of the clay and the technique of treating the surface of the ceramic fragments are very similar to the pottery from "Choga II."

As indicators of the date of the above workshop, we might point out the structure of the furnace, the exterior features of the iron slag (the thick greyish patina, a high proportion of metal), the clay plaster (ash-grey in colour), and the similarity to the ceramic produce of "Choga II," "Choga III" and "Choga 1." All this attests the simultaneous existence of all these sites. In this case, the chronological period in which they functioned cannot be determined more exactly than the 10th-8th centuries B.C.

Judging by the excavated sites, the principal distinctive feature of the Khobi-Ochkhomuri production centre is the use of haematite ores and the appearance in the workshops of special areas (trenches) for the preliminary roasting of the ore. As stated above (Chapters I-III), such trenches are totally absent from the iron-producing centres in Central and Southern Colchis where the raw material employed was magnetite sand.

Unfortunately, owing to inaccessibility of peripheral groups of iron-smelting workshops of the Supsa-Gubaseuli centre located in the headwaters of the Natanebi River, we have not yet succeeded in excavating workshops which, as we conjecture, used as raw material the ore from the poly-metallic deposits at Vakidjvari-Korisbude.

RESULTS OF LABORATORY ANALYSES OF SAMPLES FROM VARIOUS MONUMENTS IN COLCHIS (all the dates refer to the periods B.C.)

Name of site and No. of sample	Age by C^{14} analysis Laboratory of Tbilisi State University	Calibrated C^{14} dates according to P.E. Dayman *et al.*	Archaeomagnetic & Tb. Univ. Lab.dates (Z.N. Chelidze & G.I. Hutte)	Location of the sample

A. Iron-smelting furnaces

Djikhandjuri I, 1	-	-	658±100	Iron-smelting Furnace
Djikhandjuri I, 2	-	-	801±100	- do -
Djikhandjuri I, 3	-	-	620±100	- do -
Djikhandjuri II, 1	-	-	-	- do -
Djikhandjuri III, 1	-	-	-	- do -
Djikhandjuri IV	-	-	1026±100	- do -
Leğva I, 1	-	-	620±100	- do -
Leğva I, 2	-	-	818±100	- do -
Tsetskhlauri I	-	-	968±100	- do -
Tsetskhlauri II	-	-	802±100	- do -
Tsetskhlauri III	-	-	-	- do -
Charnali I, 1	-	-	705±100	- do -
Charnali I, 2	-	-	658±100	- do -
Charnali II, 1 TB-286	2670±50=720	2750±102=800	570±100	From furnace bottom
Charnali II, 2 TB-287	2720±50=770	2811±102=861	695±100	- do -
Charnali III, 1 TB-288	2750±50=800	2846±102=896	907±100	- do -
Charnali III, 2	-	-	944±100	Iron-smelting furnace
Choga I	-	-	-	-
Choga II TB-49	2596±45=646	2660±102=710	607±100	From furnace bottom
Choga III	-	-	829±100	Iron-smelting furnace
Askana I	-	-	1092±100	- do -
Askana II, 1 TB-234	3175±45=1225	3370±125=1420	1320±100	From furnace bottom
Askana II, 2 TB-235	3080±45=1130	3262±125=1312	1295±100	- do -
Askana III, 1	-	-	-	-
Askana III, 2	-	-	-	-
Askana III, 3	-	-	-	-
Askana IV	-	-	1263±100	-
Mziani I	-	-	-	-
Mziani II TB-401	2525±50=575	2578±102=628	887±100	Iron-smelting furnace 30 cm deep
Mziani II TB-402	2890±50=940	3022±80=1072	-	Same furnace 50 cm deep
Mziani II TB-403	3230±50=1280	3445±125=1495	-	Same furnace
Mziani III TB-404	2535±50-585	2588±102=638	-	Iron-smelting furnace 80 cm deep
Mziani III TB-405	2850±50-900	2970±80=1020	-	Same furnace, 90 cm deep

Mziani IV TB-407	2510±50=560	2258±102=608	804±100	Iron-smelting furnace 30 cm deep
Mziani IV TB-408	3170±45=1220	3365±125=1415	-	Same furnace bottom
Mshvidobauri I TB-431	-	-	1008±100	Iron-smelting furnace
Mshvidobauri II TB-432	3483±125=1533	3760±103=1810	-	Iron-smelting furnace from bottom
Mshvidobauri II TB-433	3270±50=1320	3491±125=1541	1207±100	Same furnace 50 cm deep
Mshvidobauri III	-	-	-	-
Mshvidobauri IV TB-434	-	-	-	-
Mshvidobauri IV TB-435	-	-	-	-
Mshvidobauri IV TB-436	3191±45=1241	3390±125=1440	1180±100	Iron-smelting furnace from bottom
Nagomari I, 1 TB-437	2780±35=830	2888±8=938	760±100	- do -
Nagomari I, 2 TB-438	3410±50=1460	3671±103=1721	1377±100	- do -
Nagomari I, 2 TB-439	3450±50=1500	3724±103=1774	-	Bottom of same furnace
Nagomari I, 2 TB-440	-	-	-	-

B. Settlements

Namcheduri III TB-319	2710±40=760	2801±102=851	-	S-E sector sq 9, depth 2 metres
Namcheduri IV TB-50	2853±45-903	2997±80=1047	-	S-E sector sq 17, depth 3 metres
Namcheduri IV TB-63	2879±50=929	3011±80=1061	-	- do -
Namcheduri V TB-320	2960±45=1010	3110±80=1160	-	S-E sector sq 26, depth 3.5 metres
Namcheduri V TB-321	2890±50=940	3022±80=1072	Tb. Lab 3000 yrs.	S-E sector sq 26, depth 4 metres
Namcheduri V TB-332	2915±50=965	3053±80=1103	-	S-E sector sq 17, depth 5 metres
Namcheduri V TB-324	3000±60=1040	3163±125=1213	-	S-E sector sq 2, depth 8 metres
Namcheduri VI TB-81	3208±50=1250	3416±125-1466	-	S-E sector sq 2, depth 10 metres
Namcheduri VI TB-230	3130±45=1180	3321±125=1371	-	S-E sector sq 17, depth 6 metres
Namcheduri VI TB-306	3440±40=1490	3708±103=1758	Tb. Lab 3370 yrs.	S-E sector sq 25, depth 4.5 metres
Namcheduri VI TB-323	3345±50=1395	3586±103=1636	-	S-E sector sq 11, depth 7 metres

CHAPTER V

RAW MATERIAL BASE AND TECHNIQUES OF IRON PRODUCTION IN ANCIENT COLCHIS

RAW MATERIAL RESOURCES

Ore

The results of research done during the past quarter of a century have clearly shown that when the ancient population of the Eastern Black Sea littoral began to master the technology of iron production they had at their disposal resources of iron ore, rich for the time in question, and the exploitation of which began at a rather early period, in fact no later than the first half of the second millennium B.C.

On the territory of Georgia there are numerous manifestations of the presence of ores and modest deposits of iron ore. In Eastern Georgia, "in the Bolnisi ore-bearing district, lies the Poladauri group of haematite deposits in a thick Upper Cretaceous volcanogenous stratum. Some of these deposits ware exploited in the past. The Adjaro-Trialeti zone contains the Dzami infiltrated magnetitic deposits connected with a palaeogenic intrusion of gabbro-diorite. Along the southern part of the Black Sea littoral there stretches a band of magnetitic sands, especially saturated with iron near the village of Supsa."[1] In fact, at the sources of the river Supsa, Natanebi, Gubazeuli and others there is a polymetallic deposit of ore, near Vakidjvarl-Korisbudé (Ozurgeti district, Georgian SSR), where massifs of ore denuded at the level of the ancient surface are, naturally, easily eroded by water, and the sand is carried down to the sea shelf where it is accumulated and distributed along the beaches. This process which began in the remote past, is still going on. It should be noted that such deposits of magnetitic sands occur under similar conditions all over the world.

From written sources we learned that peoples of the South-Eastern Black Sea littoral, the Chalybes in particular, had used since ancient times sand washed down by rivers for producing high-grade steel. In the time of Sultan Suleiman the Great, the same sands were used by the Ottoman Turks and are still used today by the Turkish government for producing iron. According to T.A. Wertime, whose expedition investigated the region in the 1960s, magnetitic sands extend along the entire southern coast of the Black Sea and westward as far as Macedonia.[2]

On the territory of ancient Colchis, magnetitic sand extended along the Eastern Black Sea littoral, from Gonio as far as Gagra. According to G. A. Tvalchrelidze, the mean proportion of magnetite along the Supsa-Natanebi beaches, at the depth of 5-6 metres, is up to 2 per cent., while the proportion of metal in concentrated magnetitic sands is about 55 per cent. In some places the proportion of magnetite is still higher.[3] In the 1950s, N.V. Khoshtaria investigated the seaside settlements of Ureki (Ozurgeti district) situated on the Black Sea terrace at a distance of a few kilometres from the coasts, as well as remnants of ancient iron-smelting furnaces; she voiced the conjecture that "the source of ore [*sic*] for the above furnaces might have been the magnetite contained, undoubtedly, in the sand of the rivers Natanebi and Bzhuzhi (with tributaries of the latter), which can be proved without any difficulty by simply approaching a magnet to the sand."[4]

Evaluating the general picture in the Colchian seaboard, and taking into consideration the history of study of the dune settlements, N.V. Khoshtaria conjectured that these settlements were connected with the use of magnetitie sand. In her words, "it is very likely that magnetitic iron obtained from sand was the first iron in this region to be forged. The high quality of magnetitic iron, its presence on the surface, and, additionally, in granules, must have freed the population from the extenuating labour of obtaining the ore from mines."[5]

Subsequently, I.A. Gzelishvili studied the problem in greater detail and came to the conclusion that "not far from the ancient metallurgical centre on the territories of the Tsetskhlauri, Djikhandjuri, Laituri and Aneseuli state farms is the richest deposit of magnetitic sand on the Black Sea coast between the rivers Supsa and Natanebi."[6] As he puts it, "despite the most painstaking investigations, no lumps of iron ore have been discovered in the waste dumps. Obviously, in those parts it was magnetitic sand that was used as ore for smelting iron; no trace of it survived in the places where iron had been smelted."[7] We might add that we also failed to find any trace of haematite or any other ore in the waste dumps, except in the iron-smelting workshops of the Khobi-Ochkhomuri iron-smelting centre, 100-120 km. away from the seashore where we found lumps of haematite in the slag dumps and discovered special trenches, open

[1] *Soviet Georgian Encyclopaedia*, "Georgian SSR," Tbilisi, 1981, p. 12.

[2] Wertheim, "The Pyrotechnological Background." *The Coming of the Iron Age*. New Haven, 1980, pp. 7, 18-19.

[3] *Mineral Resources of the Georgian SSR*, Tbilisi, 1935, pp. 244-257. *The Natural Resources of the Georgian SSR*, vol.I, Moscow, 1958, p. 83.

[4] N.V. Khoshtaria, "Archaeological Investigation of Ureki." *Materials and Research into the Archaeology of Georgia and the Caucasus*, vol.I, Tbilisi, 1955, pp. 72-73.

[5] Khoshtaria. *op. cit.*, p. 52.

[6] I.A. Gzelishvili, *Iron Metallurgy in Ancient Georgia*, p. 39.

[7] Ibid., p. 52.

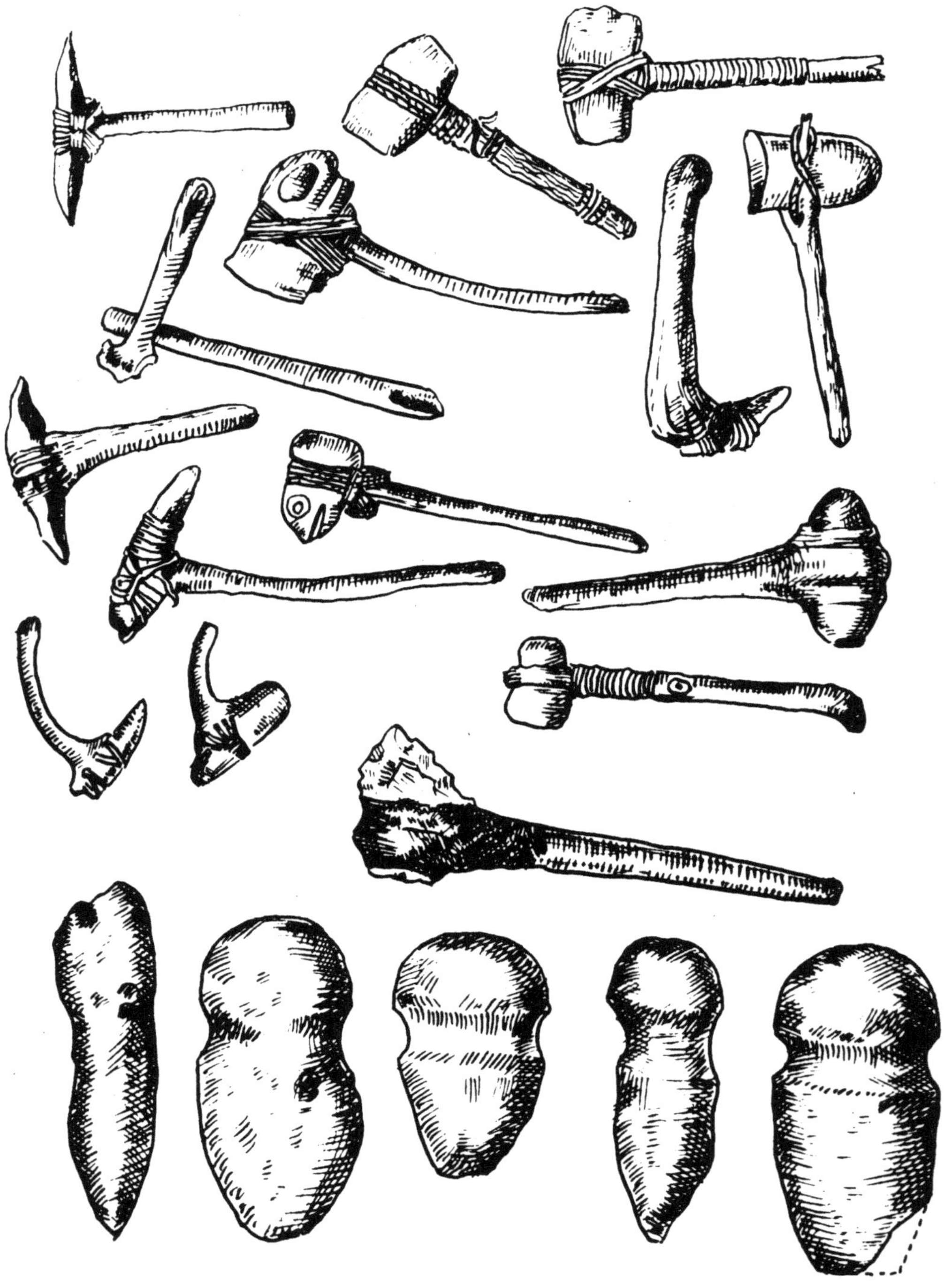

Fig. 63. Alpine miners' tools. second millennium B.C.
(after E.N. Chernykh, *Metal – Man – Time*, Moscow, 1972).

places to roast lumps of ore on (v. above, Ch.IV).[8]

At the beginning of the 1960s, we supported the opinion of N.V. Khoshtaria and I.A. Gzelishvili who stated that the ancient settlements on the seaside dunes (on the new Black Sea terrace) engaged in obtaining magnetite for iron production. Further studies of this ancient Colchian centre of iron manufacture corroborated the conclusion concerning the economic unity of the seaside settlements and the iron-producing centre situated at a distance of 10-12 km. from the seashore, between the Choloki and Ochkhomuri rivers.[9]

In the same period (1960s), A.T. Ramishvili, having made a special study of materials from the seaside settlements at Pichvnari-Kobuleti and Chakvi came to the conclusion that the thick-walled bath-tub-shaped vessels of rough clay with an admixture of sand, vessels typical

[8] Khakhutaishvili. *Materials on the History*..., pp. 3-37.

[9] D.A. Khakhutaishvili, *At the Source of Colchian Iron Metallurgy*, pp. 52-53.

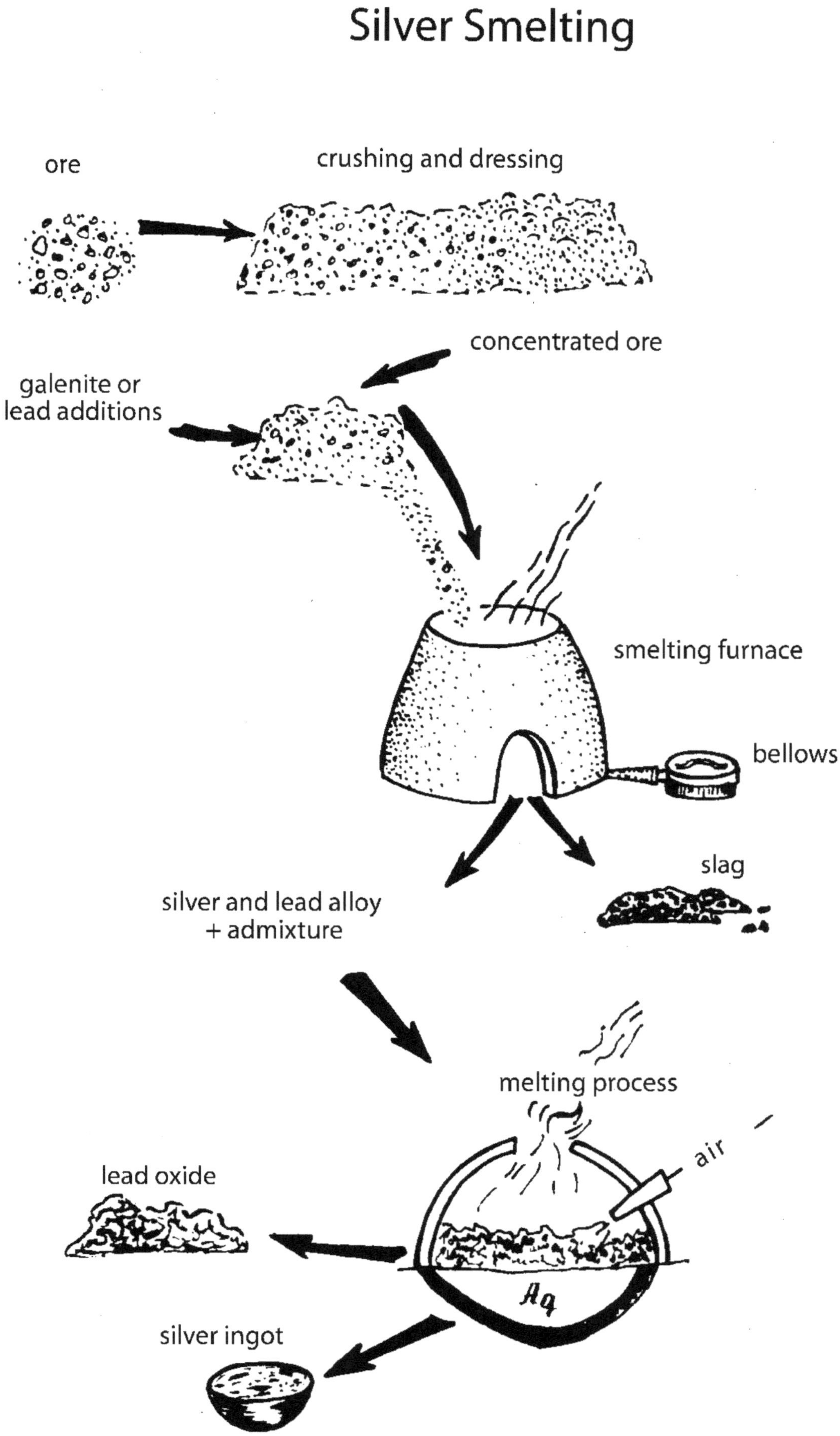

Fig. 64. Process of smelting silver (according to P.S. de Jesus)

of these settlements were intended for the flotation of magnetitic sand.[10] Subsequently he published a special work devoted to the results of his research and his reflections (see above in the Introduction).[11] During excavation of the seaside sites, he succeeded in establishing different types of bath-tub-shaped vessels (100×50×50 cm., 50×20×15 cm.), in singling out themost widespread variant (50×20×15 cm.), in pointing out miniature vessels devoid of any practical purpose; he came to the conclusion that the involvement of the seaside settlements in obtaining raw material from magnetitic sands appeared to be more likely.[12]

Subsequent research carried out by us on the territory of the Eastern Black Sea littoral, particularly in the zone between the Supsa and the Choloki rivers, confirmed that the iron-producing sites of the Colchian mining and metallurgical centre functioning in the vicinity of the seashore (the Chorokhi, Choloki-Ochkomuri and Supsa and Gubazeuli sites, as well as, possibly, the Enguri centre) used as raw material magnetitic sand from the seashore. Every iron-producing site on the shore was connected with a settlement in whose material culture the most important objects were thick-walled, bath-tub-shaped vessels that A.T. Ramishvili rightly believes were used for concentrating magnetitic sands.

Research done in the 1970s by the personnel of the Institute of Metallurgy, Academy of Sciences of the Georgian SSR and of the Mechanobrchermet Institute with the aim of discovering technically and economically the most rational methods of concentrating magnetitic sands from the Black Sea coast proved that these sands are a medium-grained material (size of grains from 0.03 to 0.5 mm.). The composition of magnetitic sand is: feldspar, quartz, fragments of shale, mica (biotite, muscovite), carbonite, pyroxene, magnetite, ferruginated particles of stone, hydroxides, as well as a small percentage of epidote, hornblende, ilmenite, argillaceous schist and apatite.[13] Mineralogical analysis has shown that the principal minerals of beach sands are magnetite and titano-magnetite (about 4.1%) as well as hydroxide of iron and ferruginated minerals (about 3.2%, rich (0.7%) and poor (0.8%) accretions,[14] and also small quantities of granules of vanadium and chrome.[15]

When studying the slag of ancient seaside centres of iron production in Colchis, C.N. Sakvarelidze and G.V. Inanishvili found titanium and vanadium. They concluded that the presence of these elements in iron slag, in industrial slags indicates the fact that the sands of this region were used as raw material in metallurgy.[16]

According to these authors who have studied the slags from the iron-smelting workshops in the Choroki, Choloki-Ochkhomuri and Supsa-Gubazeuli centres, the determining link in the composition of the slags is fayalite ($2FeO.SiO_2$), a mineral formed by iron oxide and silicon. Besides these oxides, slag also contains aluminium oxide and calcium oxide; the proportion of these oxides varies within a broad range: iron oxide 9.20-46.85%; silicon oxide 21.10-42.06%; aluminium oxide 4.60-10.85%; calcium oxide 1.20-4.07%; magnesium oxide 0.84-1.90%; manganese oxide 0.18-0.65%; and titanium oxide 0.20-1.30%.[17]

Studying the temperature of smelting in ancient bloomery furnaces, the authors found it to be between 1150 and 1250°C., corresponding approximately to the eutectic temperature of the system $FeO\text{-}Al_2O_3SiO_2$. From this fact the authors deduce that the range of fusion temperature in most samples of slag varies within 160°-190°C., which aided the rapid separation of slag and bloom in the smelting process and the formation of spongy iron with a small content of slag inclusions. This greatly facilitated the further process of forging the bloom into the half-finished product.[18] Besides, they consider it possible to state that in bloomery furnaces the temperature regimen of reduction changed step by step: the process of roasting ended before the temperature reached 1000°C., the formation of slag and the development of reduction proceeded between 1150-1250°C, while the bloom was formed and liquid slag was separated at a temperature of between 1250 and 1350°C. When such a regimen of work was kept up the iron was forged with relative ease.[19]

The authors consider that in antiquity dressing of magnetitic sands was effected by flotation. This was first suggested by N.V. Khoshtaria and substantiated by A.T. Ramishvili.[20]

As mentioned above, the use of haematite as a raw material was established in iron-producing cities away from the seashore and near the surface manifestation of iron ore deposits (the Khobi-Ochkhomuri centre, the Vakidjvari-Korisbude iron-producing site, the Supsa-Gubazeuli centre and elsewhere). Iron ore deposits occur here and there, mainly in two metalliferous regions of Georgia; in particular, in the Racha-Svaneti ore-bearing zone (the Khobi-Ochkhomuri iron-producing centre) and in the Adjaro-Trialeti mountain system (the Vakidjvari-Korisbudé site) composed of young deposits and rich in

[10] A.T. Ramishvili, "Ancient Settlements of the Sea Shore Zone in the Chakvi district," *Materials of Archaeological Survey*, 1960. *Proceedings of the Batumi Research Institute*, III, Tbilisi, 1964, p. 13 (in Georgian, summary in Russian).

[11] A.T. Ramishvili, "On the Purpose of Sites with 'Textile Ceramics' on the Eastern Black Sea Littoral." *Soviet Archaeology*, 1974, No.4, pp. 36-44.

[12] Ibid., p. 44.

[13] T.I. Sigua, A.V. Litovka, M.A. Kekelidze, "Investigation of Density of Magnetitic Sands of the Black Sea Coast." In: *Processing of Iron and Manganese ores in Transcaucasia*. Tbilisi, 1975, pp. 22-23.

[14] Ibid., p. 23.

[15] Ibid., p. 24.

[16] G.V. Inanishvili, T.N. Sakvarelidze. *Some Technological Peculiarities of Iron Metallurgy in Ancient Colchis*. 1986.

[17] Ibid.

[18] Ibid.

[19] Ibid.

[20] A.T. Ramishvili, *On the Purpose of Sites with Textile Ceramics in Eastern Black Sea Littoral*, pp. 36, 44.

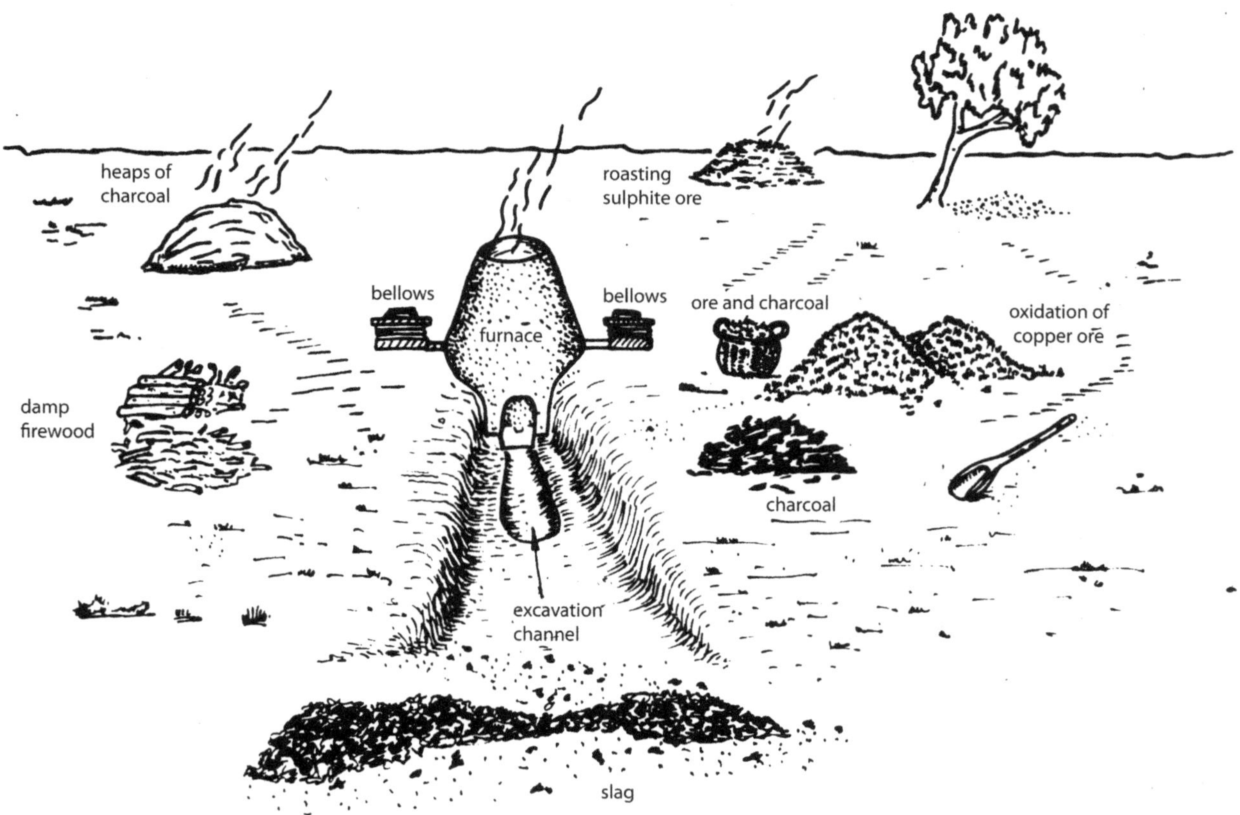

Fig. 65. Reconstruction of the Process of Metal Smelting According to the Data from Gisarsikay (after P.S. de Jesus).

manifestations of various ores. The ore deposits of the Adjaro-Trialeti system occur in the Dzami, Zekari, Gurian and Adjarian ore-bearing regions. The Gurian ore-bearing region is characterized by thick volcanic polymetallic Eocene deposits that have no practical value according to today's requirements. Here on the Vakidjvari site are magnetitic ores, while 15 km away lies the Korisbude site with copper ores. As for the Adjaria ore-bearing zone whose geophysical structurewas similar to the above it is also polymetallic, but more promising. This ore-bearing zone has considerable deposits of sulphopyrites, as well as copper and franklinite.[21]

In the ore-bearing regions of Western Transcaucasus there are deposits of magnetite, haematite, brown haematite, red magnetic haematite, etc. (Abkhazian ore-bearing region, the zone of the Georgian block, the Adjaro-Trialeti ore-bearing region).[22]

Some parts of the ore-bearing regions were developed rather early, which is attested by data obtained during the last few years as a result of studying old mine workings in Western Georgia.[23] In highland Racha alone, near the village of Gebi, over 100 sites of mining and processing ores of non-ferrous metals were discovered; sites belong to the period between the beginning of the second millennium B.C. to the first quarter of the first millennium B.C.[24] On the basis of copious material collected from the territory of Georgia, T.P. Mudjiri thought it possible to single out two centres of Bronze Age mining and metallurgy: the Greater Caucasus and Minor Caucasus. The Greater Caucasus centre, in the zone of Georgia, consists of the Abkhazian, Svanetian, Racha and Kakhetian areas; in the Minor Caucasus - the Chorokhi-Adjarian and Bolnisi-Alaverdi areas.[25] Thanks to the efforts of our scientists, Racha and Abkhazia are the more thoroughly investigated areas of ancient mining and metallurgy of all of these.

As T.P. Mudjiri puts it, "non-ferrous metal production in Racha and Abkhazia comes within the period between the second millennium B.C., and the first quarter of the first millennium B.C.," and may be subdivided into three stages:

- lst stage: the lst quarter of the second millennium B.C. to the 15th century B.C.;

[21] *Natural Resources of the Georgian SSR*, vol.I, M., 1958, pp. 15-30.
[22] Ibid., pp. 31-87.
[23] T.P. Mudjiri, M.V. Kvirikadze, "Field Investigation of Ancient Georgian Mines of the Bronze Age (Highland Racha, Highland Abkhazia)." In: *History of Mining Science and Engineering*, Tbilisi, 1979, pp. 79-82.

[24] T.P. Mudjiri, "Analysis of Regularities in the Development of Techniques and Technology of Georgian Mines in the Bronze Age." In: *History of Science*, Tbilisi, 1984, pp. 104-112.
[25] Ibid., pp. 105-106.

- 2nd stage: the 15th, 14th, 12th, 11th centuries B.C.;
- 3rd stage: the 12th, 11th, tenth, ninth centuries B.C.

He pointed out the peculiar features of each stage and showed the progressive development of mining techniques in the Middle and Late Bronze Age.[26] Thus, the development of mining techniques of the Bronze Age created all the prerequisites for their application in obtaining raw materials for iron production. Some authors suppose that while searching for various kinds of copper ores, metallurgists also found iron ores which were used as flux to be mixed with copper ore to promote the extraction of copper from it.[27] According to T. Wertheim, "the most common fluxing raw material was haematite (Fe_2O_3), the basic factor in the process of reduction of copper and lead."[28] If it was so, we have every right to state that in Colchis iron metallurgy developed from non-ferrous metallurgy.

However, the availability of ores was not the only basic condition for the inception and development of iron metallurgy.

b. Fuel

Charcoal was the principal kind of fuel used for smelting iron by ancient master-smelters.

As V.V. Ivanov remarked, "by the second half of the third millennium B.C., cardinal successes had been achieved in the selection of fuel for furnaces, and in constructing furnaces (in particular, for smelting metals) with blast force, such as is known from later Egyptian paintings. All the technical prerequisites for preparing the Iron Age had thus been created."[29]

Work carried out in the Basque Country (Vasconia) to study methods of obtaining iron in a blast furnace proved that the smelting of 100 kg. of iron required about 312 kg of ore and 340 kg. of charcoal.[30] It may be surmised that such a ratio between the raw material and the fuel existed in ancient times too.

Judging by ethnographic materials collected in Western Georgia, charcoal burning for metal smelting required special knowledge and was done by charcoal burners ("nakhshiris mokhele") who selected the required firewood, the required parts of a tree, of the trunk and branches, prepared them for burning, then sorted the results and sent them to their destination.[31]

Touching upon the significance of fire for the development of Man, Wertheim remarked that "even charcoal itself begins to acquire its particular quality only after it has been treated with fire and thus 'baked,' and when we think it has already lost its basic quality, its energy boosts up again rapidly."[32]

From the materials yielded by excavations of ancient Colchian iron-smelting workshops, we may conclude that the temperature potential achieved by master-smelters of the region reached an optimum level and, in some unregulated cases, reached the point of iron liquefaction. This was naturally furthered by a wide assortment of charcoal. In the forests of Colchis there was an abundance of trees of every kind, both coniferous and deciduous. The foothill zone of the Eastern and South-Eastern Black Sea littoral was overgrown with forests of such trees as rhododendron, boxwood, beech, cherry-laurel, alder, pine, holly, hornbeam, chestnut, cornel-tree, oak, and ash. Of these, boxwood, cornel-tree, rhododendron, oak, beech and hornbeam burned to "hard charcoal;" chestnut, pine end ash "medium-hard" charcoal; walnut, alder, and holly "soft" charcoal.[33]

Judging by ethnographic data, a specific part of a tree was used to get the required assortment of charcoal. In one case, it was the trunk, in another it was the branches that gave the required kind of charcoal. In addition, the wood that was chosen was to be of medium dampness; damp wood burned slowly; if it was dry it burnt too fast. In both cases, the final product (charcoal) was poor in quality and insufficient in quantity. The trees to be burnt were felled in August and September, when the period of growth was over.

In order to obtain charcoal, a round or square pit of the required capacity (2.5×1.5×2.0 metres) was dug; its walls were, in some cases, lined with stone. The pit had an inlet and two narrow slits to let in oxygen and let out gases. "Charcoal pits" were made either open or closed with a vaulted ceiling. Open pits were less complicated as far as structure is concerned, but they required more material: to get the same quantity of charcoal, they took twice as much wood as closed pits, and the charcoal was inferior. When the pit was charged, the fire was kindled at the bottom and when the wood burnt low some more was added again from the store that was in readiness. Regulating the process of burning needed special knowledge and skill. The intensity of the fire was regulated by blowing oxygen through the inlet and the two orifices. One cycle of burning took twenty-four hours, At the end of the procedure a specific quantity of wet crushed charcoal and earth was thrown through the

[26] Ibid., pp. 109-111.

[27] T.S. Wheeler and R. Maddin, *Metallurgy and Ancient Man*, p. 15.

[28] Wertheim, "The Pyrotechnological Background," pp. 16-17.

[29] V.V. Ivanov, *History of Slavonic and Baltic Names of Metals*, p. 37.

[30] Ibid., p. 46, with reference to B. Neumann-Wilsdorf. "Die ältesten Verfahren der Erzeugung technischen Eines durch direkte Reduction von Eisen mit Holzkohle in Rennfeneren und Stuckofen und die Stahlerzeugung unmittelbar aus dem Eisenerz. Mit einem Nachwort von H. Wilsdorf" *Freiberger Forschungshefte. Kultur und Technik*, D6, Berlin, 1954, pp. 30-31.

[31] N. M. Rekhviashvili, *Metal Forging in Racha*. Tbilisi, 1953, pp. 30-40. (in Georgian). Idem *Folk Metallurgy in Georgia*. Tbilisi, 1964.

[32] Wertheim. "The Pyrotechnological Background," p. 9.

[33] Cf. Rekhviashvili. *op. cit.*, pp. 33-34.

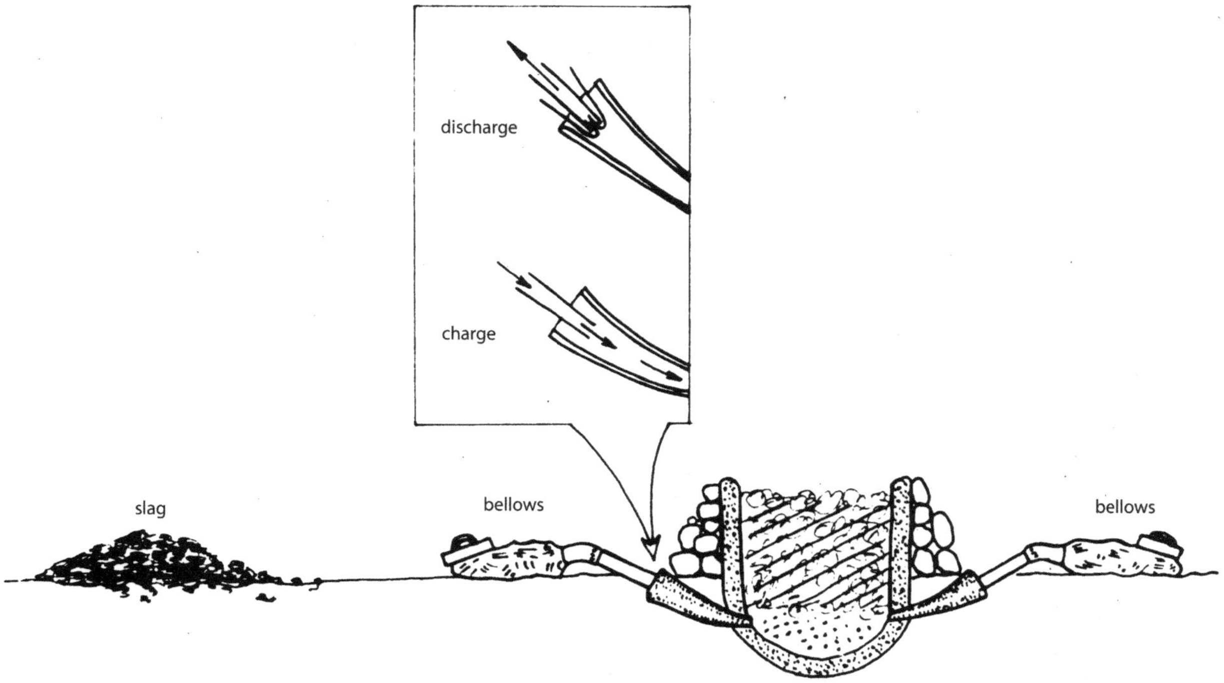

Fig. 66. Reconstruction of the Process of Metal Smelting According to the Data from Subashi (after P.S. de Jesus).

flue into the burning pit, after which the orifices were closed; then water was poured over the burning mass through the inlet which was then plastered up, stopping the burning, i.e. fully extinguishing the fire and preventing the charcoal from burning to ash. The produce thus obtained cooled off for 3-4 days, after which it was ready for use.[34] The next thing was to make the charcoal marketable, i.e. it was to be sorted by sifting through special sieves. According to archaeological and ethnographical data, ever since ancient times great attention was paid to the quality of charcoal and a simple but quite effective technology had been evolved. Without purecharcoal, it was impossible to obtain the level of temperature required for iron smelting.[35] The technology of charcoal burning had been brought to such perfection that the calorific value of the charcoal was determined by the level it occupied in the pit. The charcoal was transported by various ways: carried by men or animals on sledges or carts; it was loaded into canvas bags, in baskets of various shape and size, etc. Charcoal was stored in sheds (in ancient times) or in smithies.

During various stages of charcoal production the following tools were used: axes, saws, hammers, wedges for splitting and cutting wood; spades and shovels, picks and crowbars for digging and other earth moving work; large fine-mesh sieves for sorting out the charcoal.

We have grounds to believe that the fundamentals of the technique of producing various kinds of charcoal had been evolved at the time of the Early Bronze or even somewhat earlier, which is attested by some of the materials from Ispani, an ancient settlement buried beneath a layer of peat in an eastern suburb of the Kobuleti resort.

Fire-Clay and Refractory Clay

The abundance of fire-clay and refractory clay in Colchis was one of the main conditions for the development of iron metallurgy. It is noteworthy that the zone which in deposits of fire-clay stretches along the entire subtropic foothill area of the Greater and Minor Caucasus. In some parts of this zone there have been discovered clays particularly resistant to high temperatures, as, for instance, the deposits of fire-clay at Tsetskhlauri,[36] where the Choloki-Ochkhomuri centre of ancient iron metallurgy is situated (the Kobuleti and Ozurgeti districts). The Tsetskhlauri clays fuse at a temperature of 1750ºC. Some researchers think that the deposits of clay at Makvaneti, on the other bank of the Choloki, are still more fire-resistant.[37] In all the foothill zone of Cholchis deposits of fire-clay and refractory clay of various quality occur in other parts of the zone: at Ochkhomuri, Djinkhandjuri, Leğva, Shemokmedi, Askana, Djvarisi, Sormoni, Rioni, Khresili, Kursebé, Choga, Lia and elsewhere. Large deposits of fire-clay have been discovered at Shrosha, Tqibuli, Gelati, Tqvarcheli and elsewhere.[38]

[34] Rekhviashvili. *op. cit.*, pp. 35-37.
[35] Ibid., pp. 37-39.

[36] *Mineral Resources of the G.S.S.R.*, Tiflis, 1933, pp. 160-166.
[37] Ibid., p. 165.
[38] Ibid., pp. 146-171. M.L. Rokva, "Refractory Clays and Fire-Clays." In: *Natural Resources of the GSSR*, vol.II, Moscow, 1950, pp. 127-132.

We should note that non-industrial deposits of refractory and fire-clay occur wherever there are ancient iron-producing workshops of the Colchian centre. The ancient metallurgist thus had no trouble in obtaining the required amount of refractory or malleable fire-clay for plaster the inner surface of iron-smelting furnaces. As was mentioned above, clearing iron-smelting furnaces and waste dumps yielded numerous fragments of fire-clay plaster. Some fragments were fused to lumps of slag, which is additional proof of the conjecture that all or part of the plaster of the inner surface of the furnace had to be renewed after each cycle of smelting.

Fire-clay was also needed for making cone-shaped nozzles for bellows, many of these nozzles being required in the process of smelting metal. In some cases the nozzles were melted, deformed and filled with slag. This circumstance testifies that sometimes, at the peak of burning, the temperature in the furnace exceeded the melting point of the fire-clay at the end of the clay socket. Excavations of the ancient Colchian iron-smelting workshops showed that almost everywhere stocks of malleable fire-clay were kept under a layer of ash on the upper part of the plot where the waste dump was heaped. The clay did not dry out under the ash and preserved its plasticity.

With rare exceptions, ancient master-smelters used fire-clays of allochthonous origin. Such, for example, is the Tzetskhlauri deposit overlying tufogenic rock covered with sandstone and post-Pleiocene (Kimmeridgian stratum) conglomerate. The conglomerate was overlaid by a layer of brown clay, over which we find layers of fire-clay with charred tree-trunks embedded within them. The Tsetskhlauri refractory and fire-clays are allochthonous: they formed as a result of weathering, transportation and re-sedimentation on material of the ancient Earth's crust, represented by the bedrock of the Adjaro-Imeretian range. The deposit in question contains light-grey clay (lower horizon) and dark-grey clay (upper horizon), layers from 30 to 700 cm. thick. Of these, light-grey clay is characterized by greater plasticity, uniformity and the absence of foreign inclusions. The temperature of fusion is: maximum 1720°C., median 1680°C., minimum 1500°C. The clays from the nearby Ochkhomuri deposit have a fusion temperature between 1200 and 1660°C.[39]

Thus, ancient Colchian master-smelters, iron-smelters in particular, were provided with all the indispensable raw materials: ore, fuel and the right sorts of clay.

We know that ancient Colchian metallurgists used eutectic mixtures when smelting metals, but we cannot unfortunately, name these admixtures. The task the researchers are faced with is to re-discover and restore the picture, i.e. the flux-forming materials.

[39] Rokva, *op.cit.* pp. 134-137.

THE TECHNIQUES AND TECHNOLOGY OF PRODUCTION

The Equipment of an Iron-Smelting Workshop

There are a great many surmises and guesses about the requirements and the methods in which the technology of iron smelting was evolved,

According to the latest research, on the ancient Near East and the Eastern Black Sea littoral, iron smelting was a direct process of reduction, in which spongy iron was obtained in one operation: the smelting furnace was filled to the top with layers of charcoal and prepared ore, usually haematite and magnetite. The feeder (a clay pipe or nozzle of the bellows) was built into the masonry of the furnace, and the whole charge was ignited through the inlet for air or through the nozzle. The stream of air was first pumped in gently so as to remove the hydrate, and then with growing force so as to raise the temperature; the charcoal was thus oxidized, forming carbon monoxide which penetrated through the layers of ore and reduced it at a temperature of about 1200°C. As iron will melt at 1528°C., the metal obtained in a bloomery furnace has a spongy structure. The mass of slag composed of dirt (mostly quartz) carried off, besides metal, also part of the ferrous oxide. Inside the furnace an easily fusible silicate was formed, the greater part of which was fayalite; but fayalite is also present in the spongy mass ready for use which was then forged with a hammer to bring it up to the required condition. A considerable amount of slag was thus removed.[40]

Wertheim considers that smelted iron was first obtained by chance in furnaces meant for smelting copper and lead, in the form of slag containing a large proportion of iron; of ingots or plates. The first iron known to masters of bronze and silver production in the third millennium B.C. in Anatolia had, undoubtedly, a small carbon content. In the second millennium B.C., such iron was used for making precious objects and ornaments.[41]

He supposes that the low proportion of carbon and an abundant admixture of slag, as well as the spongy structure of iron were conditioned by the fact that before 1400 B.C. the temperature potential was low. On the other hand, iron as flux was involved in the development of a consistent technology of smelting siliceous ores of lead and iron.[42] J. Waldbaum thinks it doubtful that at the early stage of the Late Bronze period there existed an intentional smelting of iron. Even after the 13th century B.C., many texts speak for a long time only about the use of, and trade in, iron. According to V. Piggott, it was only in the latter part of the third millennium B.C. that there appeared signs that iron was manufactured with a definite purpose in view (the dagger blade from Aladja Hüyük,

[40] T.S. Wheeler, R. Maddin, *Metallurgy and Ancient Man*, p. 115.

[41] Th.A. Wertime, "The Pyrotechnological Background," p. 2.

[42] Ibid., pp.13-14.

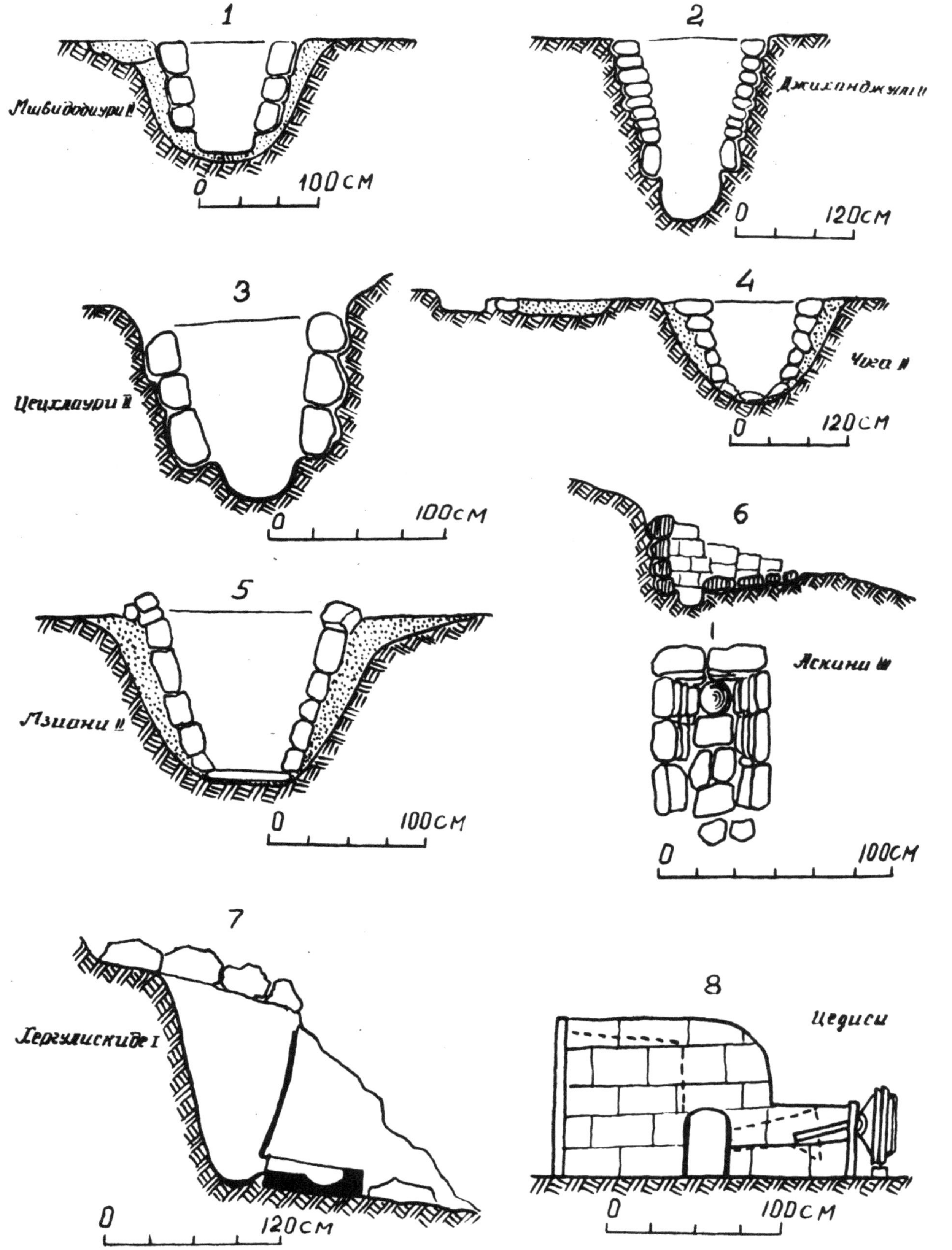

Fig. 67. Iron-smelting furnaces in Western Georgia (Eastern Black Sea Littoral). 1-5 – pre-Antique; 6 – Early Antique; 7 – Early Medieval; 8 – from the 18th-19th centuries in Racha.

Fig. 68. Iron-smelting workshop in ancient Colchis as reconstructed from archaeological data by J. Mikekadze.

dated to 2300 B.C.). He is quite right in thinking that a novelty is accepted only if it is profitable to society.[43]

Other suggestions have been made. Some specialists consider that iron was obtained in smelting copper sulphide ores, others think it resulted from the smelting of gold-bearing sands, fluxes of ores, lead, etc. (Fig. 64); but the supposition that iron could be obtained by smelting lead and copper is the most likely one.[44] Subsequently, it was proved impossible to obtain iron by smelting gold-bearing magnetitic sand, since 50% of iron is dissolved in gold at a temperature of 1100°C. Most researchers consider it possible to obtain iron used in furnaces used for smelting lead and copper (S. Smith, Th. Wertheim, I. Needham, R. Maddin, J. Muhly, T. Wheeler, V. Pigott, et al.). The same evolution might have taken place in Colchis. We have as yet no data concerning the structure of copper-smelting furnaces in ancient Colchis, but the structural similarity between Anatolian copper-smelting furnaces (Fig. 66) and the iron-smelting furnaces of ancient Colchis (Fig. 67, 1-5) does not exclude the possibility of their genetic kinship.

I.A. Gzelishvili was the first to devote special studies to the technology of iron smelting in ancient Colchis. After the excavations carried out between 1957 and 1961, this scholar had at his disposal important materials both from Western Georgia (Colchis) and from Eastern Georgia (Caucasian Iberia) which permitted him to present some well-argued statements throwing light on certain problems concerning the technical equipment of iron smelting in ancient Colchis, comparing it with materials from other regions of the ancient Near East and making a new classification of smelting furnaces.[45]

As I.A. Gzelishvili wrote, "the remains of bloomery furnaces discovered in Georgia differ in the condition in which they were found. Besides, during excavations no tools were found that are indispensable in the process of smelting. Neither vessels nor tools of wood have survived. Under such conditions, the technology of smelting iron has to be reconstructed only indirectly, resorting either to direct analogies, or to later descriptions of the processes involved which reveal the nature and regimens of work of the ancient iron-smelting workshops in which we are interested. In this respect, Georgian ethnographic material connected with iron smelting is of no small interest."[46] Having analysed numerous materials of various dates and from all the continents of the Old World, and proceeding from the data yielded by Colchian sites he had unearthed, he came to the conclusion that "in bloomery furnaces, at a temperature of 1100-1350°C. resulting from the combustion of charcoal, iron undergoes reduction in particles of ore. These tiny particles of iron together with the melted slag settle at the bottom of the furnace and, fusing together, form a spongy mass, or bloom, which is extracted from the bottom of the furnace and beaten with wooden hammers on the anvil until it is converted into a solid mass of iron good for forging."[47]

After I.A. Gzelishvili's works were published, in the 1970s-1980s, we obtained additional material which clarified a number of as yet unsolved problems of the technology of iron production in ancient Colchis. This was promoted by the discovery of new production centres and by the creation of a new, more exact chronology of not only sites of iron manufacture, but also of all the material culture of the Eastern Black Sea littoral in the third to first millennia B.C. A wider use of the achievements of the natural and technical sciences in archaeology, the excavation of a number of new contemporary settlements, making use of the most recently obtained archaeological materials from the Caucasus, the Near East and the Aegean world, a new advance of the historical sciences promoted an advance of research and studies of the equipment and gearing of metal production.

The smelting furnace

The principal element of the equipment used by master-smelters in ancient Colchis was the pit-type iron-smelting furnace[48] which did not undergo, in essence, any considerable change in structure in the course of more than 15 centuries. It was only in the sixth-fifth centuries B.C. that there appeared an iron-smelting furnace of a new structure, but in all possibility the traditional type of iron-smelting furnace continued functioning.[49]

The underground part of the earliest iron-smelting furnace "Mshvidobauri II" was a pit in the form of an overturned truncated pyramid lined with stone in the upper two-thirds of the overall height. The lower third of the pit was smeared with a thick layer of fire-clay (Fig. 49, 67, I), All the inner surface of the furnace was also plastered with fire-clay. This was done, most probably, following a tradition which was formed in the Copper and Early Bronze Ages.

Smelting furnaces of the "Mshvidobauri II" type occur in all the centres of iron manufacture, except in the Khobi-Ochkhomuri centre. In the Chorokhi and Ochkhomuri centres, all the iron-smelting furnaces are of the type described above (see Chapters I and II above); future field work may provide some corrections to this statement.

---

43 V. Pigott, "The Innovation of Iron," *Expedition*, 25, No.1, 1982.

44 Wertheim, ibid., pp. 14-15. R. Maddin, J. Muhly and T.S. Wheeler, *How the Iron Age Began*, pp. 123-124.

45 I.A. Gzelishvili, *Iron Smelting in Ancient Georgia*, p. 21 ff.; ibid., "Classification of Blooming Furnaces," *Proceedings, Tbilisi Historical and Ethnographic Museum*, I, Tbilisi, 1966, pp. 165-187, (in Georgian, summary in English).

46 Gzelishvili, ibid, p. 22.

47 Gzelishvili, *Classification of Bloomery Furnaces*., p.180.

48 Ibid., p. 165.

49 According to Tylecote, "whatever explanation may be offered, it is the smelting furnaces without a device for removing the slag that were typical of the Early Iron Age." R. F. Tylecote, *Furnaces, Crucibles and Slags*, p. 211.

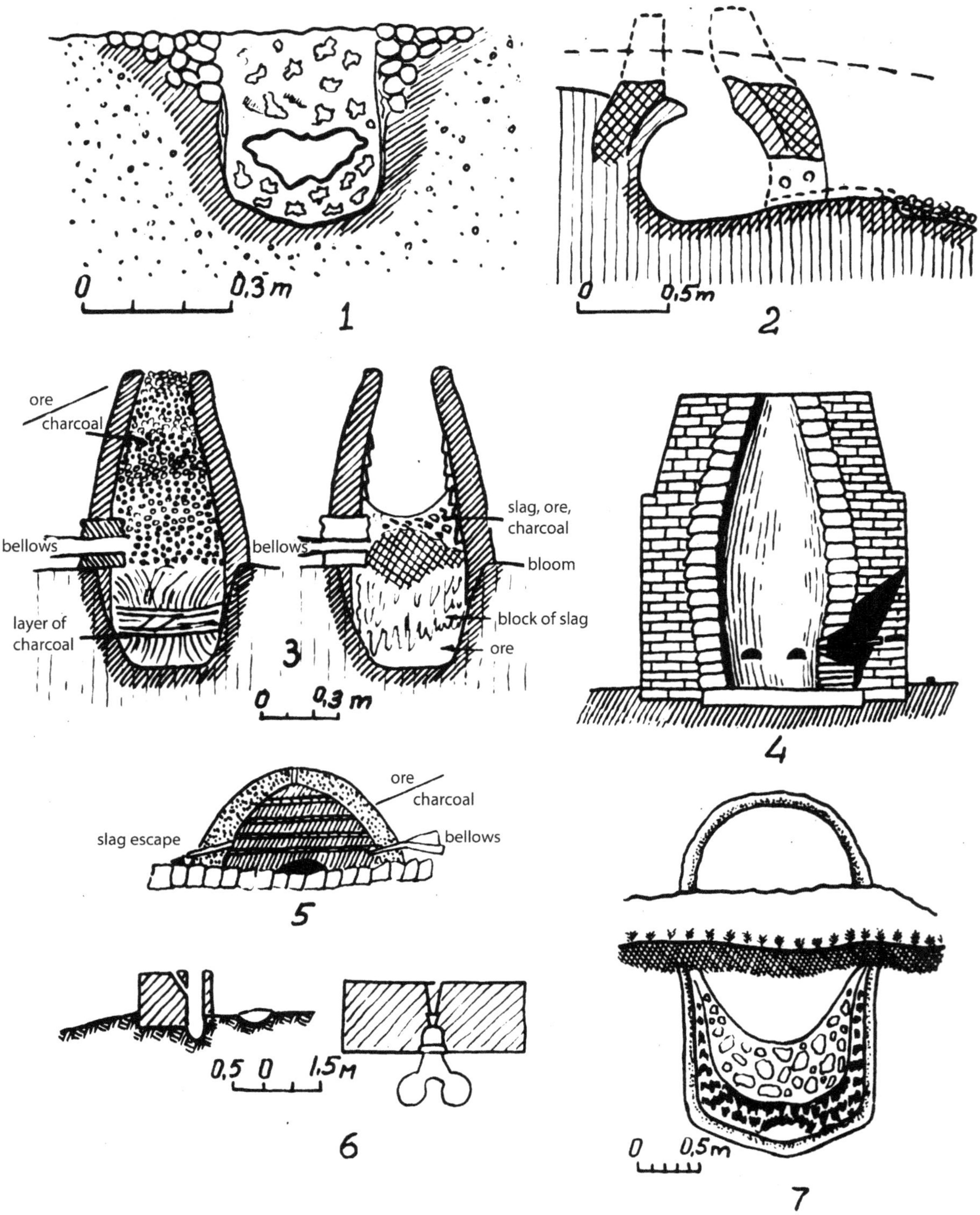

Fig. 69. Types of iron-smelting furnaces from different regions.

1) From Waschenberg, Austria. Late Hallstatt (after P. Pleiner).
2) From Silberquell. La Tène Period (after P. Pleiner).
3) Shaft furnace before and after smelting (after P. Pleiner).
4) Stückofen (after P.I. Forbes).
5) Bloomery furnace (after P.I. Forbes).
6) Pre-antique furnace from Kvemo Bolnisi (after I.A. Gzelishvili).
7) Remains of an iron-smelting furnace from Scythian times at Gorodisché (after B.A. Shramko).

The second type of smelting furnace has been registered in Northern Colchis (the Khobi-Ochkhomuri centre), in the village of Pirveli Choga; all the underground part of the furnace is lined with stone, and the bottom is hemispherical in profile (see Chapter IV). So far, we have no data about how widespread such kind of furnace was in Northern Colchis, but we can state that Choga-type furnaces are considerably younger than those from Mshvidobauri.

The third kind of furnace was identified during the excavation of the Supsa-Gubazeuli metallurgical centre. The underground part of the smelting furnaces of this kind is lined with stone, and at the bottom there is a single slab of stone. Furnaces structured in this way were first discovered on the land of Mziani village (formerly Pichkhis Djvari) in 1983. Actually, furnaces of the Mziani group do not differ, as far as structure is concerned, from furnaces of the Choga group, but according to certaindata, the Mziani furnaces are several centuries older than the former. On the other hand, it cannot be asserted that the structural peculiarities are a matter of locality. Judging by the unearthed workshop of the Supsa-Gubazeuli centre, smelting-furnaces of various types could co-exist within one centre (see Chapter III).

The superstructure of an ancient Colchian bloomery furnace was built of stone and was very likely cone-shaped (Fig. 68). Traces of the base of the conical "cap" of the furnace were identified in 1984 when the "Nagomari I, 2" site was cleared (Fig. 53, Chapter III).

In the above-ground part of the furnace, there were two projecting orifices: one in the upper part of the cone, functioning as the flue; the other, at ground level, for feeding damp oxygen (air) into the furnace through a system of blast pipes (Fig. 68); additional quantities of fuel (charcoal) were fed into the furnace through an orifice at the top.

The pit for building the furnace was dug in a thick layer of literite. When the underground part of the furnace was being lined, special "nests" were made to fit each stone used in the process of building (Fig. 69).

In furnaces with such a structure, a temperature of 1450°C. could sometimes be achieved; but, basically, the maximum temperature in these furnaces varied from 1100 to 1250°C.[50]

Research done by Georgian specialists has shown that inside the working space of the bloomery furnace there was a gradation of temperatures:

1) the roasting zone: 1100°C.
2) the zone where slag is formed and the reduction processes take place: 1150-1280°C.
3) the zone where the bloom is formed and liquefied slag is removed: 1250 - 1350°C.

Such a regimen of smelting produced iron with reasonably good forging properties.[51]

Bellows. The earliest iron-smelting bloomery furnaces in Colchis worked with air being fed into it by bellows which consisted of two principal parts: A) accordion-like bellows and B) clay nozzles (Fig. 68). The bellows were most probably made of skins and could last a long time. In the ethnographic reality of modern Western Georgia, two main kinds of bellows are known: local, and those imported from Russia.[52] The earlier are the Georgian bellows, originating locally in the Early Bronze period and found in Ispani, a settlement unearthed under a layer of peat (Fig. 70).

The nozzles of the bellows (made of clay) are conical in shape, their end widening into a funnel. These nozzles are of four types:

1) flute-like nozzles fixed inside the furnace, the side opening feeding oxygen into the furnace when the opening of the nozzle became blocked with the metal mass. (Figs. 14, 3.22, 22, 2, 3.28, 15, 17, 18.32, 4.34, 38.3, 43.6)
2) nozzles with a trench for fixing the air feedpipe consisting of several nozzles in the mouth of the leather bellows (Figs. 29.2, 42, 2, 3)
3) nozzles curved to effect the transition from air feedpipe into the sloping inner space of the furnace (Figs. 29.2, 32.5, 51.2, 4, 8, 54, 3, 60, 2)
4) simple cone-shaped nozzles — the most numerous type

The composite table was close to the furnace (Pl. I.1) and served several purposes, the principal one being used for working the spongy mass with a mallet to bring it into the required condition (see Fig. 68 showing a reconstruction of an ancient Colchian iron-smelting workshop, according to archaeological data). All bloomery smelting workshops had this element; however, not all of them have survived.

Tables were also used to prepare the charge and fuel, although in some cases a plot paved with fragments of furnace plaster and bordered with stones was arranged close to the table to store a heap of broken up ore.
A light, open shed stood at a short distance from the furnace. All the movable property of the master-smelter and his assistants was kept there. In bad weather the shed

[50] I.A. Gzelishvili, *Iron Smelting...*, p. 114, appendix II. Cf. G. V. Inanishvili, Th.N. Tavadze, T.N. Sakvarelidze, T.N. Zague, "Investigation of Ancient Slags of Iron Production on Georgian Territory." In: *History of Science*, Tbilisi, 1984, p. 21. pl.I (note: Choga belongs to the Chkhorotskhu district, not to the Batumi district as mentioned in the text in this reference).

[51] G.V. Inanishvili, T. N. Sakvarelidze, *Some Technological Peculiarities of Iron Smelting in Ancient Colchis.* Cf. Th.N. Tavadze, G.V. Inashvili, T. N. Sakvarelidze, T.N. Zague, *Investigations of Ancient Slags...*, p. 25.

[52] N.B. Rekhviashvili, *Metal Forging in Racha.* Tbilisi, 1953, p. 88 ff (in Georgian).

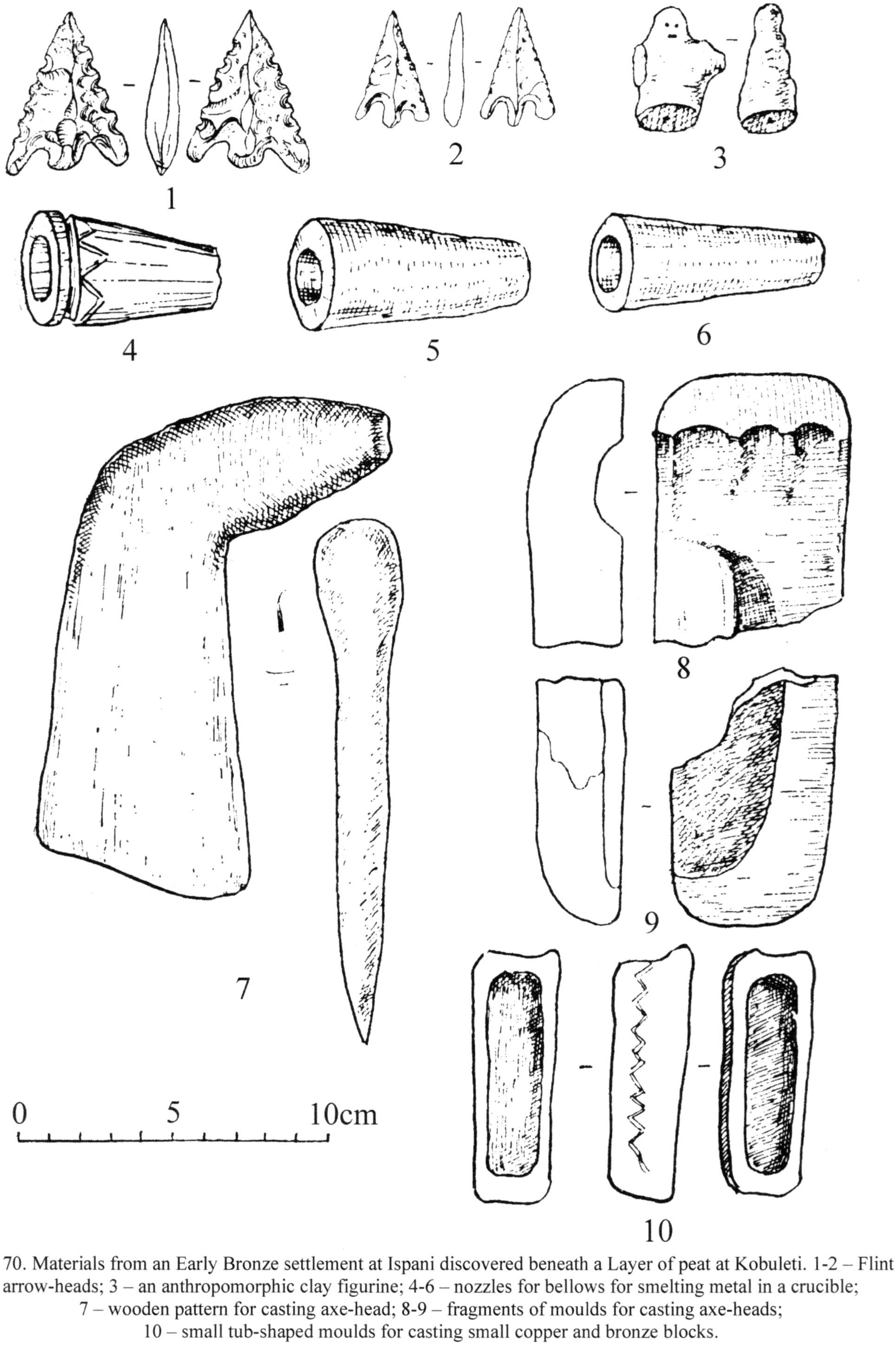

Fig. 70. Materials from an Early Bronze settlement at Ispani discovered beneath a Layer of peat at Kobuleti. 1-2 – Flint arrow-heads; 3 – an anthropomorphic clay figurine; 4-6 – nozzles for bellows for smelting metal in a crucible; 7 – wooden pattern for casting axe-head; 8-9 – fragments of moulds for casting axe-heads; 10 – small tub-shaped moulds for casting small copper and bronze blocks.

served as a shelter from rain or snow. It was roofed with shingle or thatched with rushes ("isli").

Few tools have been found, but the special operations required for smelting iron cannot be performed without special equipment for extracting the hot mass of metal and slag from the furnace, as well as for forging the bloom and forming the ingots (various tongues, wooden mallets, metal and stone hammers, various mortars and pestles for preparing flux, for crushing ore, etc.). Some of these (hammers, mortars, pestles of diorite) were unearthed during excavations; as for other tools, we can envisage retrospectively, with the aid of ethnographic data or medieval information.

The Technology of Smelting

We have no written data on the technology of iron smelting in ancient Colchis besides those provided by Pseudo-Aristotle; as we have mentioned above, from these writings we can glean information about the most ancient techniques of obtaining iron. As this anonymous writer states, the Chalybean and Amyssi iron "is obtained from sand borne by rivers; this sand, as some people say, is simply washed and smelted; according to others, the mass obtained by crushing is again washed several times and then smelted, with an addition of the so-called refractory stone in which the country abounds. This kind of iron is much better than other kinds, and if it were not smelted in the same furnace, it would not differ from silver in any way. As it is said, this iron does not corrode, but only small quantities are obtained."[53] We would be justified in conjecturing that the writer means furnaces identical to those iron-smelting furnaces of pre-Antique Colchis we have unearthed. This is all the more likely since Chalybis was not far distant from the Chorokhi centre of iron metallurgy.

We have few steel (iron) objects produced by artisans, but thanks to the materials that have survived (smelting furnaces, clay nozzles for bellows, mortars, pestles, stone hammers, open patches for roasting ore, composite tables (anvils), quantities of iron slag and plaster from the furnaces, etc.), one can easily visualize the main stages of producing spongy iron.

Problems of smelting iron in bloomery furnaces in ancient Georgia have been dealt with in the works of the late I.A. Gzelishvili. As he maintains, in Georgia "the smelting of ore was effected in small-size furnaces which were charged with ore and charcoal. The charcoal was combusted with the help of a natural draught of air or with the air blown into it via a nozzle and bellows. Iron was then reduced in particles of ore, and at the same time easily fusible highly ferrous slags appended. The grains of hard sand thus formed settled at the bottom of the furnace and, sticking together, created a spongy mass called "bloom." At the end of the smelting process this mass of iron was extracted from the bottom of the furnace with a pair of tongs and was forged with wooden mallets to solidify the iron and remove the slag from its pores."[54]

Proceeding from the achievements of contemporary science up to the 1960s concerning the essence of iron smelting in bloomery furnaces, I.A. Gzelishvili thought that the following rules were observed in ancient Georgia: in order to remove volatile substances, the ore was roasted before smelting and crushed into powder to improve the conditions for the direct reduction of iron; sulphur-rich ore was avoided (pyrites), since sulphur, combining with the metal, affected its quality. Within the furnace the temperature must be not below 1100°C., and at the end of the air feed pipe 1300-1500°C. Under such conditions the reduction of iron from the ore proceeded in a medium of carbon monoxide resulting from the combustion of charcoal in a high temperature. At the same time, part of the iron oxide transformed into pure iron which settled at the bottom of the furnace (D.Kh.), while the other part joined various oxides ("dirt"), forming slags still retaining some 30 to 40 per cent. of iron. In order to obtain steel, it would be necessary "to create such a regimen of smelting that the incandescent iron should stay longer in the medium of incandescent charcoal when the iron absorbs carbon (from 0.1 to 1 per cent.) producing soft or hard steel."[55] As seen from the above quotation (and rendering), I.A. Gzelishvili supposed, as did his predecessors,[56] that the bloom was settled at the bottom of the furnace; but a number of contemporary researchers, having conducted relevant experiments, came to other conclusions. In particular, as R. Tylecote says, "the most primitive kind of furnace for producing iron is the pit furnace, which is a simple pit in the ground or rock, into which air could be fed with the help of bellows through an air feed pipe and a low vaulted superstructure of clay. Ore and charcoal have to be mixed together or arranged in layers. The maximum temperatures must not be below 1150°C. This kind of furnace has no slag escape and it sinks to the bottom, forming a spongy mass or 'furnace bottom'. The bloom remains above the slag. After the process of smelting is terminated, the upper part of the furnace is dismantled, the bloom is taken out and the furnace is cleared."[57]

Experiments by B.A. Kolchin and O.Yu. Krug, who made a physical modelling of the bloomery process of iron production, showed that in some cases "the liquid slag sinks to the bottom of the furnace. The remaining part of ferric oxide joining with carbon monoxide becomes metallic iron. Tiny hard particles of iron descend to the lower part of the furnace as the charcoal is

[53] Aristotle. *Concerning Unbelievable Rumours*, 830. In: T.S Kaukhchishvili. *Greek Authors about Georgia*, II, pp. 67-68., ibid., *Ancient Greek Sources of Georgian History*, p. 68 (in Georgian).

[54] I.A. Gzelishvili, *Iron Metallurgy in Ancient Georgia.*

[55] Ibid., p. 21-22.

[56] See: Voskoboinikov, "A Description of Mining and Useful Minerals in the Regions of Imereti, Racha, Mingrelia, Lechkhumi and Suaneti." *Mining Journal*, St. Petersburg, 1826, vol. 11, pp. 51-58.

[57] R. Tylecote, *Furnaces, Crucibles and Slag*, pp. 210-211.

consumed, forming spongy iron permeated with slag."[58] In other cases, the experiment yielded smallish bits of pure spongy iron and a large lump of metal conglomerate inside which there were also bits of spongy iron mixed with slag. Pure pieces of spongy iron were above the lump of conglomerate.[59]

Although the process was modelled in a Russian smelting furnace that had a slag escape, the consecutive order of the chemical processes involved in smelting was, most probably, not very different from what we can observe in pit smelting furnaces without a slag escape.[60]

As a result of an investigation of bloomery slags from ancient Colchis, it was found that the presence of the same type of slags (chemical and mineralogical composition) in nearly all the sites of iron-smelting industry is a convincing proof of the similar condition in which it was obtained and the use of smelting furnaces of a similar structure. There probably was a gradation of temperatures in the working space of the furnaces:

1) the zone of roasting: up to 1000°C.
2) the zone where slag was formed and the process of reduction proceeded: 1150-1250°C.
3) the zone where the bloom was formed and liquid flowing slag was discharged: 1250-1350°C.

these conditions ensured the production of iron and steel with a sufficiently stable malleability."[61]

Judging by the latest data of laboratory investigations of ancient slags and manufacture of iron, the ancient Colchian artisans had accumulated a great experience in the technology of smelting. It has been possible to determine that "the temperature interval for the softening of most samples is between 160 and 190°C, which furthered the rapid separation of the slag from the bloom in the smelting process and the obtaining of spongy iron with a small quantity of slag inclusions. This circumstance facilitated the further handling of the bloom to obtain half-made objects."[62]

In iron-smelting furnaces of the pit type which have no slag escape, to which the Colchian furnaces belong, the sinking of the melted substance proceeded layer after layer, for the various components of the charge melted at different temperatures. First, easily fusible substances sank, while refractory substances overlaid them. As B.A. Kolchin says, "in blooming furnaces without slag escape, the process resulted in the formation of a lump of metallurgical conglomerate. It consisted of slag, spongy iron, and remnants of charcoal. The iron was in the top portion of the lump, the slag in its lower portion. In order to obtain pure iron, another operation was required to separate the slag. The iron could be separated by mechanically breaking up the lump, picking out the iron and welding it together again, or by smelting the lump of conglomerate in special vessels."[63]

In the foothills of Colchis, the zone in which the remnants of iron-smelting workshops are scattered, even without archaeological excavations, one can find and collect on the waste dumps very rich material for a detailed study of as yet unsolved problems of the technology of smelting metals, of iron in particular. The melting temperatures of samples of slag collected from the unearthed objects is between 1150 and 1250°C., and the fire-clay plaster of the furnace withstood a temperature above 1450°C.[64] Analysis of the chemical composition of ancient slags (Addendum 1) gives us a hope that other problems of this ancient technology will soon be solved.

Judging by archaeological materials, the Colchian iron-smelters made a wide use of various fluxes, as attested by data provided by analyses of slags, as well as by numerous mortars and pestles for crushing various kinds of materials. Forging the hot bloom played an important part in the process of obtaining marketable iron; this was done on the composite table arranged close to the smelting furnace, with a wooden mallet.[65]

Abundant iron dust around the stone anvil shows that part of the metal was lost when the spongy mass was forged. Contemporary experiments in physical modelling of the smelting processes in a bloomery furnace carried out both in this country and abroad have demonstrated that the production of marketable iron in such a furnace required a most meticulous approach, a substantial knowledge of the peculiarities of separate elements determining a successful smelting (the ore, kinds of charcoal, fluxes, regulating the blast, etc.). As B.A. Kolchin puts it, "the process of producing marketable blooms, i.e. welding 5-6 kg of iron (the maximum weight of ancient Russian bloom) was a rather complicated and difficult matter. It was technically impossible to weld a large lump of spongy iron into a monolithic lump of pure metal. It was necessary to bring the iron many times to high temperatures, subsequently forging it vigorously and repeatedly, skilfully using various fluxes which liquefied the slags that impregnated the sponge. This must have been a separate labour-consuming arduous technological

[58] B.A. Kolchin, O.Yu. Krug, "Physical Modelling of the Bloomery Process in Iron Production." In: *Archaeology and the Natural Sciences*, Moscow, 1965, p. 208.

[59] Ibid., p. 209-214: "Reduced spongy iron does not sink to the very bottom of the furnace, but stops somewhat higher. At the bottom there is always liquid slag."

[60] See: Tylecote, *op.cit.*, pp. 209-223. Wertheim, "The Pyrotechnologic Background," pp. 12-20.

[61] G.V. Inanishvili, T.N. Sakvarelidze, *Some Technological Features of Iron Smelting in Ancient Georgia*, p. 6. Cf. Th. N. Tavadze *et al.*, "Investigation of Ancient Slags of Iron Production on Georgian Territory." *History of Science*, Tbilisi, 1984, p. 24.

[62] Ibid., p. 6. Cf. Tavadze et al., *op.cit.*, p. 25.

[63] B.A. Kolchin, O.Yu. Krug, *op.cit.*, p. 215.

[64] I.A. Gzelishvili. *Iron Smelting in Ancient Georgia*, p. 114, Addendum II.

[65] Ibid., pp. 25-26.

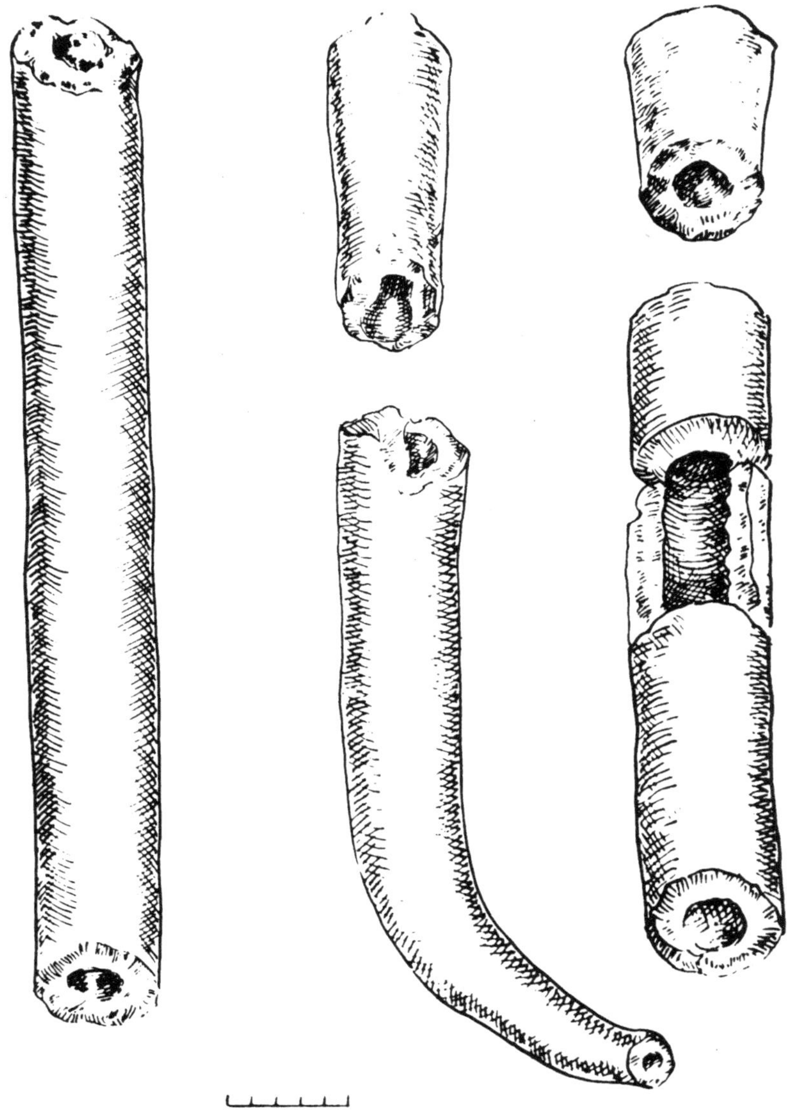

Fig. 71. Clay nozzles for bellows for crucible process, from ancient Colchian settlement "Namcheduri V."

operation which could be performed both by metallurgists and blacksmiths."[66]

Thus, we may state that as far back as the first half of the second millennium B.C. this "separate labour-consuming, arduous technological operation" was exclusively the job of metal workers who provided marketable metal for the emerging iron market.

On the Organization of Labour

Problems of the organization of metal production, of the social status of people occupied in industry, the social and political anatomy of a society that used a highly developed technology of metal smelting, etc. all this is most important to an understanding of the historical processes taking place within Colchian society in the second and lst millennia B.C.

Judging by the letter of Hattusili III to Salmanasar I, iron smelting in ancient Anatolia (and perhaps in the regions immediately bordering on to it) was not practised all the year round, but only seasonally.[67] Naturally, the communal structure of society in early pre-class and class stages gives no grounds to expect absolute specialization, complete division of labour among the members of a community, or between communities; however, according to data from ancient written sources (Xenophon's *Anabasis* and *Argonautica* by Apollonius of Rhodes *et al.*), in the classical epoch, in the zone of the metal-producing regions of the Black Sea littoral, in particular in the region that was the home country of

[66] B.A. Kolchin, O.Yu. Krug, "Physical Modelling of the Blooming Processes in Iron Production." In: *Archaeology and the Natural Sciences*, Moscow, 1965, p. 209.

[67] G.G. Giorgadze, *Production and Use of Iron in Central Anatolia According to Data from Hittite Cuneiform Texts*, p,16. In the opinion of O.R. Gurney, the stock of iron ran short with the Hittites basically by the end of summer, as the peasants engaged in metal smelting only in winter when they were free of agricultural work. *The Hittites*, Harmondsworth, 1961. pp. 83-84; cf. E. Menabde. *Hittite Society*. Tbilisi, 1965, note 34.

Kartvelian (Georgian) tribes, there seemed to have existed communities (or tribes) of professional iron workers. We might suggest that all these communities of metal workers were not founded on kinship, but on a territorial basis. The Hittites used to grant plots of land to artisans and probably let them settle in the regions more favourable for industry.[68] On the other hand, the traditional skills of the artisans cannot be disregarded, and entire communities were formed of craftsmen specializing in a certain trade.

The seasonal character of work in iron production, the fact that it was done when the worker was free from tilling his land and doing other field work, i.e. in winter, is corroborated by ethnographical examples.[69]

According to N. Voskoboinikov, A. Petzholdt, N. Karpinsky, N. Rekhviashvili, et al., folk metallurgy in the highlands of Western Georgia, and in Racha in particular, had preserved certain features that are deeply rooted in the past. N. Voskoboinikov says that in Racha, the peasants in the village of Tsedisi smelted iron and made various implements from it. The nearby ore deposit was worked by hand, with picks and crowbars, not blasted with gunpowder;...they engage in this handicraft only when they are free of agricultural work or other work about the farm, which occurs in autumn and in winter."[70]

When N. Voskoboinikov visited this village, it had 19 iron-smelting furnaces working with the help of leather bellows with clay nozzles. The process of smelting required three men: the master-smelter and two assistants attending to the bellows.[71]

A. Petzholdt was astounded when he found himself near 'a metallurgical plant.' "It was a building square in plan, each side 14 feet long. 10 feet high, in which not only the principal process took place, i.e. producing the bloom, but also working it to obtain forging iron for making iron tools. I could not believe my eyes that all these processes took place in such a small plant. At the back wall (in the same single room) instead of a 'high furnace' was the bloomery furnace built of rough stone blocks...."[72] All this equipment was put into operation by hand-bellows. "When one gets acquainted with this work, one cannot believe that bloom is obtained here, but the result proves it. The bloom is not yet pure metal; it contains a lot of slag. The final smelting contained both bloom and forging iron. I must stress that with the needed skill such metal is used for making various iron implements: picks, axes, spades, etc.). In this plant, at the right wall, we also saw a 'cooling furnace' where forging iron is produced (previously such a furnace was meant for a smithy). It is built of stone reinforced by wooden boards and plastered with clay. On the surface it appears as a pit where the iron is cooled with the help of bellows operated as mentioned above. When the iron is 'cooled,' it is moved to an anvil in the middle of the plant where the required implements are forged from it. In the mornings a man comes to the plant, with a basket of ore, and leaves the place late in the evening with a pick (or another implement) that he has made." When viewing the deposits of iron ore, Petzholdt became convinced that "work was done here mainly in winter.. and each man carried away as much as he required for his personal needs."[73]

Of course, Petzholdt was not aware that the population of Racha, a West-Georgian province,[74] were direct descendants of the ancient Colchians, and the seemingly primitive way in which the Rachians obtained iron was based upon a simple technology of smelting iron in bloomery furnaces evolved in the course of centuries back in the Bronze Age. The highly effective simplicity that so surprised Petzholdt was the particular value of this technological achievement. In this connection, it is not without interest to mention that in Old Georgian, as well as in the languages of several Caucasian peoples, steel was denoted by the term "çanari," i.e. Chanian.[75] The Chans were one of the groups of Colchian (Kartvelian) tribes to which the so-called Chalybes also belonged.

Before serfdom was abolished in Georgia (in 1864), Tsedisi village belonged to the Tseretelis, an Imeretian princely family; the peasants, naturally owed their master the corvée, as well as a tax paid in iron and iron goods, which often led to violent social conflicts.[76]

We have no concrete data concerning the organization of labour in Colchian iron metallurgy, but we may through light on the problem by means of indirect information. Thus, the temporary character of the stations on the new Black Sea terrace so rich in magnetitic sands means that this raw material was most probably stored seasonally, in a period when men were not engaged in agricultural work. i.e. in autumn and in winter. If so, it is logical to suppose that smelting iron was also done seasonally, in the colder months of the year. The seasonal nature of work with metals is justified by the fact that it is wiser to engage in a "hot" trade in winter than in summer. Collecting all the materials required for smelting iron (the ore, fire-clay, building the furnace, modelling the nozzles for the bellows, making the leather bellows, preparing the

[68] Menabde, *Hittite Society*, p. 66, ref.34.

[69] See for example: K.N. Serbina, *Peasant Iron Metallurgy in North-Western Russia from the 16th to the First Half of the 19th Century*. Leningrad, 1971, pp. 95, 174, etc.

[70] N. Voskoboinikov, *op.cit.*, pp. 51-52. I.A. Gzelishvili, *op. cit.*, pp. 22-25.

[71] Ibid., p. 52. A. Petzholdt, *Der Kaukasus*. Leipzig, 1866, pp. 317-320.

[72] According to N. Voskoboinikov, in the village of Tsedisi "each furnace is served by two cylindrical stone bellows. These furnaces are of the following dimensions within: 19 inches deep, 19 inches broad at the top and 13.5 inches at the bottom. In some, in the lower part, there is a semicircular opening to let the slag excape. About 16 kg. of ore could be treated at a time, obtaining about half that weight of bloom which, treated in another furnace, yields about 4 kg. of pure iron... each smelting takes 6 hrs. and requires over 80 kg. Ibid., pp. 51-52.

[73] A. Petzholdt. Ibid., p. 320. I.A. Gzelishvili, op.cit., pp. 24-25.

[74] For details see: N.B. Rekhviashvili, *Forging Metals in Racha*, pp. 61-87.; idem., *Folk Metallurgy in Georgia* (Ethnographic essay), Tbilisi, 1964, p. 66 ff.

[75] Rekhviashvili, *Folk Metallurgy...*, p. 168 ff.

[76] Ibid., pp. 11-17.

fluxes, etc.) presupposed a certain kind of cooperation the organization of which, it may be supposed, was, before any state organisation came into being, in the hands of the community élite. There are some data attesting that since the Middle Bronze Age there were, in some advanced regions of the Transcaucasus, some powerful communities that exploited ordinary communities in various economic spheres. Most probably, forced labour is more easily exploited in handicrafts. The powerful communities that controlled the production of iron dwelt at the sources of raw materials, as the materials of the Colchian foothills show. As an example, we might cite the ancient Colchian settlement of Namcheduri[77] situated beyond the new Black Sea terrace that supplied the Choloki-Ochkhamuri iron-producing centre with magnetitic raw material (see above). Most probably, this community organized the manufacture of iron and the sale of the surplus produce.

At the beginning of the lst millennium B.C., after the rise of the Colchian Kingdom, it was very likely the royal administration that organized the production of iron and its sale on foreign markets; through the viceroys, the King controlled and regulated the output of the finished product.

In one of his inscriptions discovered in the vicinity of present-day Artanudji, on the site of the ancient Colchian town of Ildamush, King Sarduri II of Urartu narrates "I...marched on the land of Culkha: I. took by storm Ildamush; its people I burned; LUirdi of the land of Culkha who was there I put to death. I prepared an iron seal and set up an inscription in the town of Ildamush."[78] As specialists conjecture, LUirdi means, in the language of Urartu, "ruler," or "viceroy."[79]

As was noted 15 years ago, "it is very likely that viceroys or other officials were placed in authority in every region and militarily strategic, or economic, centre of Colchis. It is not excluded that metallurgical centres, as well as other objects of state significance were subordinated to the viceroys. Speaking of the viceroys of Culkha in the ancient times in the East, one may recall the institution of the 'skeptukhs' of Colchis. According to Strabo, after the sailing of the Argonauts "the succeeding kings divided the country into skeptukhiae and were thus weakened."[80] The institution of the skeptukhiae was very likely the actual institution of LUirdi ('viceroy') of the Urartian sources, or a subsequent modification of it."[81]

For lack of sufficient material, we are unable to present all the details of the organization of labour in mining and metallurgy, but we can state with certainty that all the preliminary work was done by the manufacturers themselves who had the status of dependent members of the community. At the end of the present review, we might quote Apollonius of Rhodes who speaks about the Chalybes:[82]

> *On the next day by nightfall they arrived in the land of the Chalybes.*
> *That people is not inclined to tilling the soil with oxen,*
> *Neither to grow various fruits that rejoice the soul;*
> *Nor do they graze flocks of sheep in pastures with abundant dew, the Chalybes.*
> *Hewing the iron-bearing earth, its resistant entrails,*
> *They receive a wage - that which feeds them,*
> *And no dawn rises without them being at work, -*
> *Arduous labour it is, in smoke and black soot.*

A comparison of this description by Apollonius of Rhodes with actual ethnographic reality in the West-Georgian province of Racha with regard to the population of the Sarkineti gorge[83] shows that little has changed in the course of two thousand years in the conditions under which the people work. In 1939, the ethnographer N.B. Rekhviashvili wrote down the words of Iliko Khidasheli, a resident of Racha: "Formerly the people of neighbouring villages were reluctant to give their daughters in marriage to men from Sarkineti, considering them unsuitable for fear that their daughters would be over-burdened with work: it was not unfounded, for the work of men from Sarkineti was arduous and joyless."[84]

Concerning the social status of metallurgists in antiquity, we might cite Xenophon who wrote that "the Chalybes dwelling on the coast of the Black Sea are not numerous and are subordinated to the Mossynoici, and they live mostly by obtaining and working iron."[85]

Thus, Georgia's folk metallurgy preserved, almost to the middle of the 19th century, certain features of ancient iron metallurgy.

---

[77] T.K. Mikeladze, D.A. Khakhutaishvili, *Namcheduri, an Ancient Colchian Settlement*, Tbilisi, 1985.
[78] G.A. Melikishvili, *Urartian Cuneiform Inscriptions*. Moscow, 1960, p. 304.
[79] Ibid., p. 398.
[80] T.S. Kaukhchishvili, *Strabo's Geography*. Tbilisi, 1957, p. 124.
[81] D.A. Khakhutaishvili, "On the History of Ancient Colchian Iron Metallurgy." In: *Problems of Ancient History* (Caucasian-Near-Eastern Collection, IV. Tbilisi, 1973, pp. 178-179.

[82] Apollonius of Rhodes, *Argonautica*.
[83] The term "sarkineti" derives from the Georgian word "rkina" (iron) and denotes the place where iron is produced and forged. See: N.B. Rekhviashvili. *Forging Iron in Racha*, p. 8 ff.
[84] N.B. Rekhviashvili. *op. cit.* II, ref.3.
[85] Xenophon, *Anabasis*, vol.V, Ch.V,I.

PLATES

I. 1. Tcharnali I. View from the south.
2. Tcharnali II. View from the north after excavation.

II. Tcharnali I.
1,2. Iron furnaces in the process of excavation.

III. Tcharnali III. 1. View from the south after excavation.
2. Iron smelting furnace 'Tcharnali III' view from the south

IV. Tsetskhlauri II
1. Waste heap before excavations. View from the north.
2. A heap of iron slag and fragments of clay lining from the furnaces.

V. 1. Iron smelting furnace 'Mziani II'. Viewed from above
2.'Mziani III, 1' after excavation. View from the south.

VI. 1. Nagomari I, View from the north.
2. Furnace for Iron making 'Nagomari I, 1' during archaeomagnetic investigations.

www.ingramcontent.com/pod-product-compliance
Lightning Source LLC
LaVergne TN
LVHW071218210926
848482LV00015B/843